AF566460

LINDA C. POPE

Sustainability:
A Call to Action

Part I
Sustainability at the Individual Scale

Linda C. Pope

Square Root Press
Portland, OR 97232
squarerootpress@gmail.com

First edition, BookBaby, September 2014
Second printing, March 2016

Second Edition, Square Root Press, August 2016

Cover design by Guy Boster
Cover photograph by Linda C. Pope

Print ISBN- 978-0-9977560-0-5
E-book ISBN- 978-0-9977560-1-2

DEDICATION

For anyone trying to make a difference,
may you find the courage
to rise to the occasion,
and be brilliant.

For my parents, my son, and S.C.

LINDA C. POPE

PREFACE

The path towards sustainability began for me by reading 18 books on the topic. But as I finished the third one, I realized that we already had all the tools, all the technology, everything we needed to live sustainably. We just aren't doing it. So I culled the best, the most inspiring parts of those books and put together a course to inspire individual people to step up to the plate and, hopefully, become leaders in their environments. I am grateful to the authors of those books for their powerful words. This book represents a summation of those words.

Despite the power of their words, most of the books that were inspiring me were 10 to 15 years old. Surely we have made some progress since then. So wherever I was able, I researched to describe our current situation. In some cases we have made great improvement, in other cases we are still in decline. The success stories, however, show that it is possible to change. We just have to do it. I hope the message from all these authors will inspire you to also change.

Linda C. Pope

CONTENTS

ACKNOWLEDGMENTS

There are so many people that I need to thank for helping me put this book together. SC provided infinite support, helping me to express an unimaginable possibility. Two people inspired me to write: Dr. John Bonner, who at 94 had just completed his 27th book, and my son, Timothy Shiels. Timothy kept me on track every time he asked, "What about your book?" He also created the title for the book. Siarra Voller did a great job as the initial reader. Cindy Brown helped with Excel and with the construction of the graphs. Rio Hibler did a wonderful job as editor and APA formatter. Andrew Carlson was an excellent proofreader and publishing consultant. Guy Boster created the beautiful cover design. All of these people helped within the environment of "I needed it yesterday!" I appreciate the many friends that tolerated my learning curve and indulged me as I learned about the book-writing process. Thanks to my sustainability students who have always shown me that I do have something to say, and that this information is a source of inspiration for them.

CHAPTER 1
SUSTAINABILITY: WHAT IS IT?

For decades, if not centuries, we have each contributed to the current degradation of our ecosystems. Garrett Hardin (1968) explained the dilemma in his classic tale of unsustainability, *The Tragedy of the Commons*. Unless a common, in this case the pasture, is regulated in some fashion, we take advantage of and abuse the situation to our individual benefit, to the detriment of all. Individually, we will continue to add more cows to the common pasture until our standard of living declines and the ecosystem fails.

After learning more about sustainability, we understand that this does not have to be our fate. It does mean that we have to change the way we currently do *everything*. The current model of taking from the earth to make new products that are quickly sent to the landfill must be replaced with a cyclical model, in which resources are used in continuous cycles, and there is no waste. This must be the focus of our efforts for change in the next 3 to 4 decades.

We live in challenging times. Every day, we are bombarded with disheartening news and, with each situation, we have the choice to ignore the issues or ask, "What can we do to change this?" There has never been a time in which an emphasis on hope and inspiration is more important than now. Moore (2013) describes hope as a free energy source and a powerful influence that will strengthen our ability to overcome all challenges. Recognition of this energy source could not have come at a better time! I am writing this book, a condensed source of information regarding our current status, to give hope and inspiration, and to provide possible solutions for now and our future. This book is asking you to "step up to the plate" and act!

The terms "sustainable" and "sustainability" are commonplace, and they are often overused. People confuse environmental science with sustainability. Environmental Science is related to all of the earth sciences (biology, chemistry, soil science, atmospheric science, geology, ecology, geography), whereas sustainability includes the interaction between humans and nature, and economic and social aspects of fairness and equity. Learning how to live in harmony with the planet and nature is essential. If we each can work to leave this world in better shape than we currently find it, we will succeed. None of us have to do it alone. We have each other to lend support. We just need to start, to do our part no matter how small, and to keep looking for new and better ways to find the harmonious balance we seek, as well as to become a source of inspiration for others.

The most commonly quoted definition for sustainability was first described by the Brundtland Commission of the United Nations (1987): "Sustainable development is development that meets the needs of the present without compromising the ability of future generations to meet their own needs." If humankind is to survive, then embracing sustainability in all of its forms is crucial. However, even sustainability is not enough; we need to move beyond sustainability, (which "involves scarcity and minimalism") to a world that is thriving with "abundance and enrichment" (Edwards, 2010).

This is an exciting time. Those who truly engage this challenge will get to redefine the way our world functions at every level. There is almost nothing that is already perfect just as it is other than nature. There are rumblings of change everywhere. We are on the cusp of a tipping point. Because of this, it is important that we each find our piece in this puzzle that we can contribute to transform this world, and our future world will become the amazing place we all want it to be. Sustainability involves finding methods for humans to develop and grow within the bounds that nature provides. Thriving, on the other hand, includes not only sustaining the environment but also restoring it to its former health and well-being, while at the same time providing a multitude of possibilities wherein we can each express ourselves and grow.

What is science? It is the main tool we use to understand our surroundings. We use our senses and our ability to think in order to collect data that will explain what we experience. Science represents what we know as well as the process we use to gain that knowledge: the scientific method. This method is a procedure that has attempted to explain natural science for over 400 years. It consists of systematic observation, careful measurement, and repeated experiments after the formulation and modification of hypotheses (Oxford, 2013). Where pure science may seek information for its own sake, environmental science seeks to solve environmental problems.

These problems were not created intentionally, and in fact we may have been oblivious to the potential harm we were causing. The Earth is so large and humans are so small. Surely it is not possible for humans to change the Earth in any significant manner!

Why do we need to change? We are discovering that all webs of life and all food chains are interconnected. And because of the misunderstanding we have had regarding our place in the world, all ecosystems are in decline (Millennium, 2005). The current culmination of all environmental degradation is a result of people trying to survive, trying to make a living, and then those actions expanded to the point of greed. Many of these environmental issues are controversial. Sometimes they are fabricated, or clouded by special interest groups; sometimes they are caused by contradictory world views; in many cases the problems are just very complicated. Environmental science is the most complicated of all the sciences. It requires understanding many interacting disciplines: earth and life sciences; psychology, culture and sociology; history, politics and economics. It is our place in history to deal with the ramifications of centuries of abuse to nature and to each other.

Environmental science describes the natural environment and its imbalances within systems, either natural or anthropogenic (caused by humankind). Sustainability addresses how we can refrain from further degradation by changing our methods and by the use of conservation. Using principles of thrivability, we can bring systems back to health, referred to as restoration ecology. Sustainability and thrivability function at many scales, and this will be described in subsequent chapters.

Monitoring and assessing our progress along these paths of change requires a way to measure progress, or some form of an indicator. This monitoring is frequently organized in the form of "The Triple Bottom Line" (Figure 1.1). Unless all aspects of life are taken into account, nothing can be truly sustainable. As Ray Anderson, former CEO of Interface Inc. says: "You can't make a green product in a brown company, and you can't have a green company without a green supply chain" (2006). Similarly, you cannot have a green (sustainable) city if the state is brown (unsustainable), or a green country if the rest of the world is brown. By this I mean no one, no region, no time period can be degraded at the expense of creating one "perfect" location. For example, if we

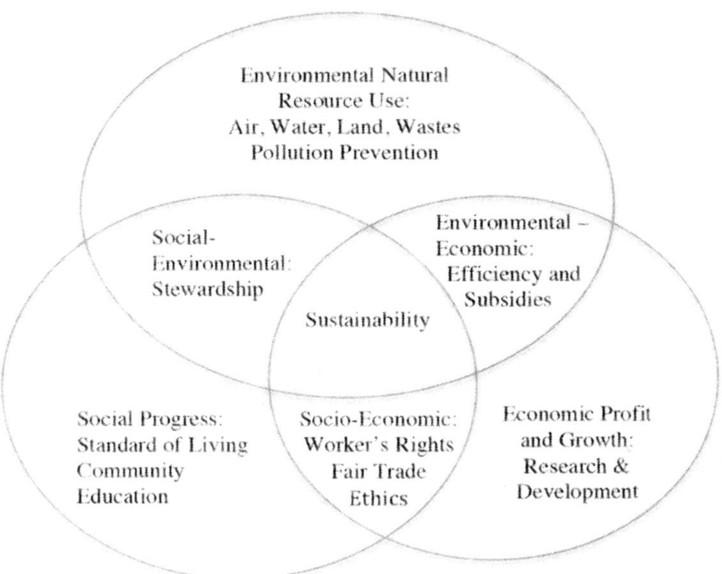

Figure 1.1. The Triple Bottom Line. Source: Adapted from the University of Michigan Sustainability Assessment (2002).

send our trash to the other side of the world, it *will* be burned there, but the toxins will enter the air stream and return to us in the form of chemical-laden rain. There is no "away."

The Triple Bottom Line (Figure 1.1) represents what needs to be measured in sustainability, but not how to measure it. The Cities Programme of the UN Global Compact has modified the traditional diagram to include a measurement indicator (Figure 1.2), the Circles of Sustainability. An indicator informs you when a condition is operating as planned; it can also indicate the direction needed to bring an issue back into balance. Indicators are varied, but they always have some characteristics in common: they must be relevant, easy to understand, reliable, and based on data that is easily accessible (Hart, 2010).

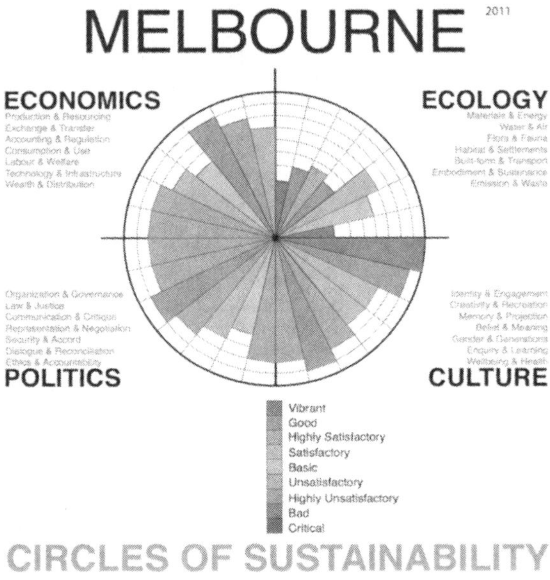

Figure 1.2. Circles of Sustainability. Source: UN Global Compact Cities Programme Secretariat, directed by Paul James.

Instead of using the Circles of Sustainability provided by the United Nations, the United States issues reports. For example, the National Climate Assessment Development Advisory Committee summarizes the concerns of the United States in a report of over 1,000 pages; individual chapters are available online (NCADAC, 2013). The executive summary is a frank description of the degradation that is currently happening in our country and what is expected for our future. The last 2 chapters address mitigation and adaptation, and these are recommended for review (NCADAC,

2013). However, the average person might better understand our current status if we used a visual graphic such the Circles of Sustainability instead.

How can we know that we are making the right choices? The best answer is to use nature as a guide. This emerging field of looking to nature for answers is called biomimicry (Benyus, 1997). For example, as temperatures increase, the mountain pine beetle is no longer killed off during the winter months. An increase in their population leads to excessive forest decline. We respond by harvesting the dead or dying trees to increase yield from the forest. When a tree dies naturally, the wood, needles, and leaves decay, and return nutrients to the soil. By harvesting this timber, we prevent the natural cycle. According to Hawken (1993), when these trees are dying they put out a noise, some may call it a song, which is heard by the beetle. This calls the beetle to feed on the tree and to turn it back into rich soil for the next generation. What if there are not enough nutrients in the soil to support healthy growth, and the tree can sense this limitation? In response, it “sacrifices itself” for the benefit of future trees, and calls the beetle so that its return to the soil will enrich the growth of future generations. If this is true, then we continually prevent the nutrients from returning to the soil. We are disrupting the natural sequence of events. Is the solution to let this generation of trees die so that the future forest health can be restored? Would it be best if we learned how to help nature to heal itself? These are questions for this generation to address.

As a species, we have been living a wild teenage life, doing whatever we want. In comparison to all life on Earth, we are a young species and have not fully grasped the consequences of our actions. We have reached a point in history where we must now mature as a species and get on a sustainable path. Oren Lyons says, “What if we choose to eradicate ourselves from this Earth, by whatever means? The Earth goes nowhere. And in time, it will regenerate, and all the lakes will be pristine. The rivers, the waters, the mountains, everything will be green again. It'll be peaceful. There may not be people, but the Earth will regenerate. And you know why? — Because the Earth has all the time in the world and we don't. So I think that's where we're at, right now” (DiCaprio, 2007). We do have another option. We can choose an alternative path, the one of sustainability. It is up to us to make this decision everyday, in every action. It will not be easy work, but the work will always be amazing.

The population of the world is over 7.2 billion people. The questions we have to ask ourselves are, “What is the carrying capacity of the earth?” and “Are we smarter than yeast?” Yeast is added to grape juice to make wine. The yeast cells multiply and consume the sugar in the juice until their numbers have expanded to excess. This continues until all the sugar is gone, and the yeast cells die. The shape of their lifespan on a graph is called “overshoot and collapse” (Meadows, 2004) (Figure 1.3). The current shape of human population growth is following the exponential growth shown by yeast as it expresses unlimited expansion (Figure 1.4). Will we control our population growth and unrestrained use of resources and bring our civilization into a carrying capacity in balance with other ecosystems, or will we follow the path of yeast in wine – overshoot and collapse? Understanding all aspects of our existence and the impact we have on each other, other cultures, and other species is clearly paramount. We have the opportunity right now to bring

Earth's ecosystems back to a natural equilibrium, to live sustainably and to reach a stable carrying capacity (Figure 1.5).

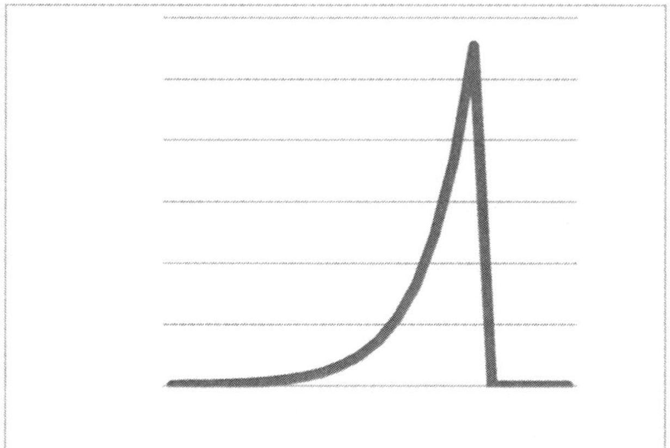

Figure. 1.3. Growth curve for yeast: Overshoot and collapse. Source: Meadows, Meadows and Randers (1992).

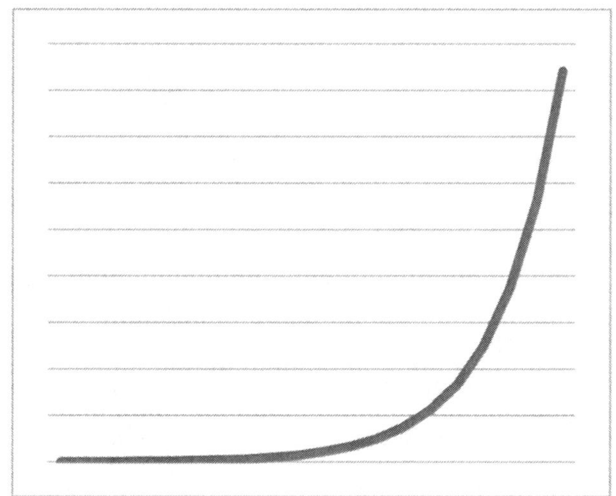

Figure 1.4. Growth curve for human population, following the exponential portion of the yeast graph (Figure 1.3).

How will we do this? Not only do we have to consider future generations, but also we must simultaneously address the well-being of all other species and cultures. Also we must develop a restorative viewpoint, not only to live sustainably, but to restore all systems to health: our forests, our agricultural land, the oceans and the wetlands, because *we* are the fire that will ignite this movement, this

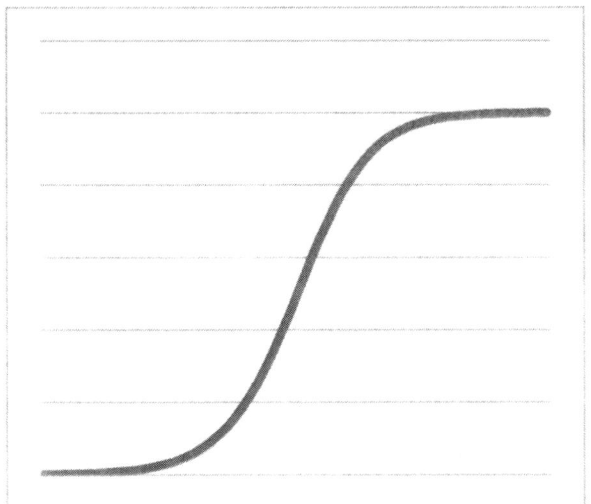

Figure 1.5. Logistic growth curve for a population that reaches a stable carrying capacity, in balance with other species in the ecosystem. Source: Thompson, 2014. Logistic Population Growth. Population Ecology, Encyclopædia Brittanica.

sustainability movement. The movement depends on us. Actually, all life on earth depends on us.

It is time to change. And changing is hard! We resist! Even if everything is going to be so much better than we can imagine once we do change, we still resist! What is the first thing you can do? According to environmentalist Bill McKibben, you can "bake cookies and take them to your neighbors and get to know each other" (Edwards, 2010). This community connection is the beginning of global connections. How can we have peace in the Middle East if we cannot even extend ourselves to our next-door neighbors? That disconnect must be repaired at the smallest scale.

We also need to value what we have left. Native Americans responded by setting aside sacred sites: a valley never to be hunted, a stream never to be fished, a grove of trees never to be cut. We created our National Parks with this idea in mind; however, for decades we have abused those lands, and have not considered them to be sacred (Smith, 1954; Butler, 2010). We have to appreciate what we can see now, for those sites may not be available for our grandchildren.

When I was a teenager living in Europe with my family, I was one of the few who got to see the real prehistoric cave paintings in Lascaux, France, before they were closed to the public. The body heat of the tourists caused a fungus to grow and obscure the magnificence of those paintings. Now, only dedicated scientists are allowed to enter the caves. I also walked with my family and dog around the stones at Stonehenge, England. The millions of visitors each year caused significant erosion and tourists are no longer permitted access or the right to actually touch the stones. When I lived in the state of Washington in the sixties, salmon runs were so intense that, under every bridge, the salmon were so plentiful that you felt you could walk on the backs of the salmon to cross the streams. What will we be denied in the future? Obviously seeing the glaciers and ice-fields at the poles will be in jeopardy, but what else? Tourists still take boat rides to the base of the glaciers and watch the face calve (break off), something we may not be able to experience for more than another few decades. To actually just see a glacier may not be possible in the future due to the warming of our planet.

The scope of the issues is frequently not within the grasp of our everyday experiences. Education helps. Every class has the opportunity to have a profound effect on the participants and their communities. For example, a permaculture class in Kinsale, Ireland (year-round population 2,257), wanted to move their community toward lower energy consumption (Kinsale, 2005). The students took into account subjects such as energy use, food production, transportation, marine resources and tourism, and created "Kinsale 2021: An Energy Descent Action Plan." This grassroots plan that aims to rebuild communities in the shadow of peak oil (the oil that was cheap and easy to extract), climate instability, and economic hardships, resulted in the "transition town initiative" and, as of the writing of this chapter, there are more than 1,095 registered initiatives following the model developed by that permaculture class (Transition, n.d.).

"Another world is possible. Let's build it" (Internetartizans, 2008)! Where do we start? The first rule of sustainability seems to be to recycle. We all know we can recycle. Some states and some cities are better at it than others. We know to change our incandescent light bulbs to compact fluorescent ones, or to LED light bulbs to save energy. We know to drive less and to walk more. Beyond these minimal changes, however, most people have no idea how deep sustainability really runs, and there are few, if any, guides to help us. Strangely enough, we can learn from business. There are dozens of sustainability models developed for the business world (Edwards, 2005; Edwards, 2010), and they can serve to guide individual actions. These models provide the basic tools needed to address the changes we need to make at the individual, community, and global levels.

Edwards describes many models that are used by commerce, communities, natural resources, etc. (2005; 2010). *The Sustainability*

Street Approach (Edwards, 2010) was developed by Frank Red, an environmental educator. It has two unusual and important principles: "Spurn doom and gloom" and "Take baby steps." These two steps are critical to any sustainability plan, but are often overlooked. It is essential to have hope if we are to succeed (Moore, 2013), and it is not possible to do it all at once.

To illustrate the "baby steps," I moved into assisted living to help care for my elderly father. We were served our food restaurant-style, and we would get two orange juices in plastic to-go cups, twice a day. At first I was recycling the plastic cups. But there were so many! So then I started to reuse the cups. I'd wash them out, and we would carry them downstairs to be used again. But then I read about the plasticizers that were probably leaching out into my orange juice. I replaced the plastic cups with a glass container that was reusable. I then began to look at what I was putting *in* the container. Orange juice should be considered a treat, as it has to travel 3,000 miles (4,828 km) to reach me in Oregon. I switched to apple juice, which probably came from the Pacific Northwest, where I live. But apples are on the "dirty-dozen" list, which means you cannot even wash the pesticides and other chemicals off (EWG, 2013). So I switched to organic apple juice (or tomato juice) grown and packaged in Washington or Oregon in glass bottles, and transferred it to a glass container for my personal use. Many "baby" steps were needed to transition to this single sustainable action.

We have many denial mechanisms built into our patterns and vocabulary. Doppelt lists the most common:

1. "We don't have to do that." This means you do not really care;

2. "We tried it but it didn't work." Sustainability measures were probably not attempted;
3. "We already do that." This keeps the organization rooted in the linear-production model, as opposed to the cradle-to-cradle viewpoint;
4. "The successes are mostly anecdotal so we'll wait for more hard data." This is often used by the government with the idea that it is safer to follow than to lead;
5. "It's too costly (or time consuming or complicated)." Costs probably weren't calculated; and,
6. "It's --- fault, not ours." This, of course, just tries to place the blame elsewhere, and supports the view that nothing need be done. (Doppelt, 2003)

It is quite normal to feel resistance and try to avoid action. We have to acknowledge the fact that we have made a mistake, and that we are doing something harmful. And we have to beware of becoming agitated and defensive.

Perhaps you're in denial and you ask, "How can we make a difference? Maybe it is too expensive to change." Hawken et al. (1999) describes an all-glass tower in Chicago that needed to replace its leaking windows. By using Superwindows, their cooling expense was reduced 85%; the cooling equipment needed was much smaller, more efficient, and $200,000 cheaper. If all glass buildings in America did a similar renovation, $45 billion per year could be saved. Business continues to learn that becoming sustainable does not cost money, it saves it.

So. What would our future look like if all actions were 100% sustainable? In order to reach that point, we need to work to inspire everyone (including ourselves!) to develop sustainable practices in every aspect of our lives. Work sustainability into everything you do, every friend you meet. Start NOW! I recommend that you read 3 books about sustainability. They will transform you!

It won't be possible to become globally sustainable one building at a time, or one household at a time. One sustainable business or household is *great*, but that's just not fast enough. The scale of the city or state is too large, too slow, too expensive. *All scales are important*, but what we do in communities is much more than we are able or inclined to do as individuals (Seltzer, 2010). The community or neighborhood scale is the most important. We must learn what to do as individuals so that we can build a community effort. Final comments by some of the participants at the EcoDistrict Summit 2011 were that this is the decade of urgency. We have to inspire everyone in the country to develop his or her own neighborhood sustainability plan by 2020. That will leave 20 to 30 years to implement the plans. And everything must change by 2050 if we are to protect the stability of our weather.

The main idea is to start. You can't do it all at once, and it is normal to feel resistance. So, make a list to include:

1. Things you would like to change, know need to be changed, but are not possible now;
2. Things that you <u>can</u> change, however small, and work on those right away;
3. Keep adding to the list; and,
4. Keep working on #1!

The scout for the Oregon Trail (Figure 1.6), Kit Carson, said, "The cowards never started and the weak died on the way." Perhaps the same goes for the Sustainability Trail. Now is the time to really get on board and become a source of inspiration for all of those around us. This book is that call to action.

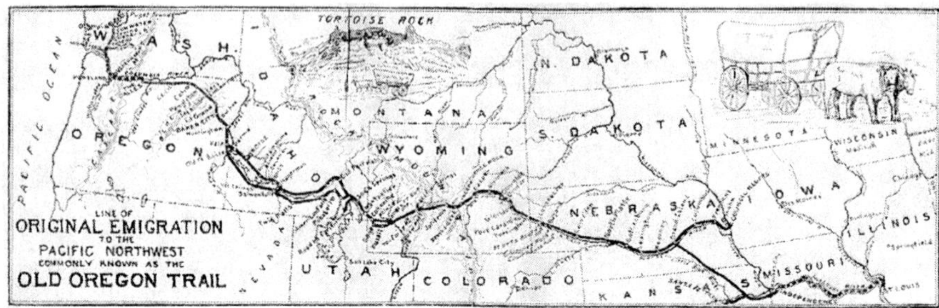

Figure 1.6. The Oregon Trail west. Line of Original Emigration to the Pacific Northwest Commonly Known as the "Old Oregon Trail." Portland, one of the most sustainable cities in the country, is the new starting point for the trail East – the Sustainability Trail! Source: The Ox Team or the Old Oregon Trail 1852-1906 by Ezra Meeker.

References Cited for Chapter 1
Sustainability: What is it?

Anderson, Ray. (2006). Restoring the Planet, One Carpet Tile at a Time: An Interview with Ray Anderson. *Green Technology Magazine*. Published online. Retrieved from http://www.green-technology.org/green_technology_magazine/ray_anderson.htm.

Benyus, Janine. (1997). *Biomimicry: Innovation Inspired by Nature*. New York, NY: HarperCollins Publishers Inc.

Brundtland Commission of the United Nations. (1987). Report of the World Commission on Environment and Development: Our Common Future. Transmitted to the General Assembly as an Annex to document A/42/427 - Development and International Co-operation: Environment. Retrieved from http://www.un-documents.net/ocf-02.htm

Butler, Katherine. (2010). National parks abused by visitors abusing technology. *Mother Nature Network*. Retrieved from http://www.mnn.com/earth-matters/wilderness-resources/stories/national-parks-abused-by-visitors-abusing-technology

Circles of Sustainability from the Global Compact, Cities Programme. (n.d.). Retrieved from http://citiesprogramme.com/archives/resource/ circles-of-sustainability-urban-profile-process

DiCaprio, Leonardo (Producer) & Petersen, Leila Conners and Conners, Nadia (Directors).(2007). *The 11th Hour*, United States: Warner Independent Pictures.

Doppelt, Bob. (2003). *Leading Change Toward Sustainability: A Change-Management Guide for Business, Government and Civil Society*. Sheffield, United Kingdom: Greenleaf Publishing Limited.

Edwards, Andres. (2005). *The Sustainability Revolution: Portrait of a Paradigm Shift*. Gabriola Island, BC, Canada: New Society Publishers.

Edwards, Andres. (2010). *Thriving Beyond Sustainability: Pathways to a Resilient Society.* Gabriola Island, BC, Canada: New Society Publishers.

Environmental Working Group. (2013). *EWG's 2013 Shopper's Guide to Pesticides in Produce™.* Retrieved from http://www.ewg.org/foodnews/summary.php

Hart, Maureen. (2010). *Sustainable Measures*. West Hartford, CT. Retrieved from http://www.sustainablemeasures.com/node/92

Hawken, Paul. (1993). *The Ecology of Commerce: A Declaration of Sustainability*. New York, NY: HarperCollins Publishers Inc.

Hawken, Paul, Amory Lovins, and L. Hunter Lovins. (1999). *Natural Capitalism*. New York, NY: Little Brown and Company.

Internetartizans. (2008). Another World is Possible. Retrieved from http://www.internetartizans.co.uk/Another_World_Is_Possible

Kinsale, 2021: An Energy Descent Action Plan. Version.1. (2005). Edited by Rob Hawkins, Kinsale Further Education College, Kinsale, Ireland. Retrieved from http://transitionculture.org/wp-content/uploads/KinsaleEnergyDescentActionPlan.pdf

Meadows, Donella, Dennis Meadows and Jorgen Randers. (2004). *Limits to Growth: The 30-year Update.* White River Junction, VT: Chelsea Green Publishing.

Millennium Ecosystem Assessment. (2005). *Ecosystems and Human Well-being: Synthesis*. Washington, DC: Island Press. Retrieved from:
http://www.unep.org/maweb/documents/document.356.aspx.pdf

Moore, Frances. (2013). *Why Sharing News About Solutions is a Revolutionary Act*. Nation of Change, Human Rights. Published online at http://www.nationofchange.org/why-sharing-news-about-solutions-revolutionary-act-1368358423

National Climate Assessment Development Advisory Committee. (2013). U.S. Global Change Research Program. Retrieved from http://ncadac.globalchange.gov/

Oxford University Press. (2013). Online version. Retrieved from http://oxforddictionaries.com/us/definition/american_english/

Seltzer, Ethan. (2010). *Introduction, Making EcoDistricts: Concepts and Methods for Advancing Sustainability in Neighborhoods, Portland, OR*. Retrieved from http://www.pdx.edu/planning-sustainability/sites/www.pdx.edu.planning-sustainability/files/Bullitt%20Foundation%20Paper%20-%20Making%20EcoDistricts%20-%20%20PSU%20Research.pdf

Smith, Susan D. (1954). Vandalism and Abuse in National Parks and Monuments, *Journal of Geography*, Volume 53, Issue 7, pp. 291-297.

Thompson, John N. (2014). Logistic Population Growth. Population Ecology, Encyclopædia Brittanica. Retrieved from http://www.britannica.com/EBchecked/topic/470416/population-ecology/70579/Logistic-population-growth

Transition Initiatives Directory. (n.d.). Retrieved from http://www.transitionnetwork.org/initiatives

World population. (n.d.). Retrieved from http://www.worldometers.info/world-population/

CHAPTER 2
IMPORTANCE OF VISION AND SUSTAINABILITY MODELS

"Pack your bags! I've won a vacation for two! I've already contacted your family and your boss. I'll pick you up in an hour." What do you pack? Where are you going? How long will you be gone? If you don't know where you are going, how will you know when you get there? Or *how* to get there, or if you are even on the right road? You have to have a plan, a *vision*.

What is a vision? If you do an Internet search for "importance of vision," innumerable examples are given describing the paramount requirement of having "a vision" in business. Jeff Weiner, CEO of LinkedIn, describes vision as "The dream; a team's true north. Primary objective is to inspire and create a shared sense of purpose throughout the company" (Weiner, 2012). More often a vision describes what you can accomplish, and for great leaders, "a vision is not seen as a dream, but a reality that has not come into existence" (Burke, 2009). "A vision brings meaning to peoples' work, mobilizes them to action, and helps them decide what to do and what not to do in the course of their work" (Evolution, 2013).

Definitions of "vision," however, do not have such motivating descriptions, the best being, "the ability to think about and plan for the future, using intelligence and imagination, especially in politics and business," or "someone's idea or hope of how something should be done, or how it will be in the future" (Macmillan, 2009-2013). Most business sources agree that if you have a vision, everything falls into place. My favorite vision was the one that Ray Anderson, former CEO of Interface Inc. gave his company: "Be the first company that, by its deeds, shows the entire world what sustainability is in all its dimensions: people, process, product, place and profits—and in doing so, become restorative through the power of influence" (Anderson, 1994). His number one rule when reaching for "Mission Zero" was, "Set an outrageously ambitious goal to inspire real change" (Interface, 2010). This is a true vision that inspires everyone to go beyond what is expected of him or her to reach true greatness.

Today, our definition of success is "How much stuff do you have?" Tomorrow, success will mean doing more with less, being more efficient. Perhaps being small and producing high-quality products and services will be more profitable than being "corporate." You cannot measure your success, or your progress, if you do not have a clear destination. Without a vision, people eventually become discouraged and give up. "Vision without action is just a dream. Action without vision just passes the time. Vision with action can change the world" (Barker, 2013). Knowing what you want to achieve requires a clear vision. Vision describes what you plan to do. If you only follow the law, comply, you strive to avoid legal action (Doppelt, 2003). That framework does not inspire greatness.

A common view described by Doppelt (2003) for vision is called the backwards-looking approach, or "backcasting." Keller (2013) describes backcasting as the "challenge for the Young Global

Leaders to design a world that works for everyone where even our most vulnerable can live with dignity and fulfillment. Creating that world is the challenge for us all."

Think about what your future would look like if it were sustainable in the next 5, 10, or 25 years. How would your life be different? How would purchases be made? Would services change? What about the way we make things? What are sources of our raw materials? Would where we live or work be different? What about our sources of energy and water? How would we deal with wastes of all kinds? What would these systems look like and how would they function in a sustainable world? Select and describe a point in an ideal future where everything, everywhere, for everyone, is completely sustainable. Then look backwards from that point and to view your current condition. What are the steps that are needed for you reach that ideal state? Where is the shortest path? What are the steps that can be accomplished right away and which steps will take more time?

Hawken et al. (1999) offer the following vision:

> Imagine for a moment a world where cities have become peaceful and serene because cars and buses are whisper quiet, vehicles exhaust only water vapor, and parks and greenways have replaced unneeded urban freeways. Living standards for all people have dramatically improved, particularly for the poor and those in developing countries. Involuntary unemployment no longer exists, and income taxes have largely been eliminated. Houses, even low income housing units, pay part of their mortgage costs by the energy they *produce*; there are few if any landfills; worldwide forest cover is increasing; dams are being dismantled; atmospheric CO_2 levels are decreasing for the first time in two hundred years; and effluent water leaving

> factories is cleaner than the water coming into them. In communities and towns, churches, corporations, and labor groups promote a new living-wage.
>
> (Hawken et al., 1999, p.1.)

This is not a utopian vision. Everything that is needed for this transformation to take place already exists. This vision just needs you to help drive it into being (Hawken et al., 1999). In the mid-18th century, before the industrial revolution, if you had predicted that the productivity of humankind would increase dramatically, and that one person would soon be able to do the work of two hundred, you would have been labeled as totally crazy. The same situation is occurring today as predictions of our resource use in production have the potential to increase five, ten or even a hundredfold. When resources were abundant, we found ways to make what we wanted; however that way was not the most efficient way. It was just *a way*. Now we will find the *best* way to make the same product, and find new ways to use less and less energy and resources in the process.

One vision for our new future is that of "zero impact." This term is not widely understood (Steffen, 2011). In general, we don't take the time to truly grasp the impact of our daily actions. Where do things really come from? Where do they really go when I no longer want them? Any idea of zero impact is truly scary, and "looms like a giant wall of deprivation" (Steffen, 2011). However, if we work at it and get it right, it really can be the good life for every single person on earth!

In order to have a sustainable world, we each need to develop our own vision statement. Every scale, from the individual to the neighborhood, the state, the nation, and the world, they all need to have a vision. Our individual vision cannot exclude the needs of any of the other scales. For example, when thinking about our wastes, there is no "away." We have to develop a vision plan that deals with our wastes where they are generated, and learn to avoid creating them in the first place. What is waste? Is it just a process that was never designed correctly?

So how do you develop a vision for a sustainable future, when your current life situation has not exposed you to the range of issues that must be considered? Where do you start? Fortunately, business has provided us with models we can follow as individuals. The effort that they have expended can now inspire us too.

One model, the Natural Step, has beautifully described the scientific basis used to create their 4-part model for business (Natural Step, n.d.):

First: "Nothing disappears," as reflected in the law of the conservation of energy and mass (First Law of Thermodynamics). Matter on earth has stayed the same for billions of years. It may change its form, but it cannot be created or destroyed. A tree takes in carbon dioxide and transforms it into wood. When the wood is burned, or the tree dies and decays, the solid matter is returned to carbon dioxide again.

Second: "Everything spreads" describes the natural tendency for matter to disperse (Second Law of Thermodynamics), or entropy. Although the total amount of energy remains the same, the amount available for use after each transformation decreases, due to entropy. Entropy, or chaos in a system, is always increasing. This means that anything that is created by our human society will gradually disperse throughout all of nature. Again, there is no "away."

Third: "There is Value in Structure," says that we place value on the actual shape of the materials. Although we use the matter that has taken this particular shape, we cannot consume the energy or matter itself. We value the service and aesthetics of our sunglasses. If we accidentally step on them, what we have valued is destroyed, but all of the atoms in the structure remain the same and have not changed.

And fourth: "Photosynthesis Pays the Bills." The energy available on the earth is almost entirely either directly or indirectly generated through photosynthesis. The chloroplasts within the plant cells capture photons of light (energy from the sun), and transform it into an energy form that can be used later either by the plant or another organism.

When describing matter on earth, it operates within a closed system, with increasing disorder. However, since there is a constant flow of energy (light) from the sun, this is an open system, continually adding new structure and order to offset the constantly increasing disorder (The Natural Step, n.d.). It is these basic scientific principles that we have been ignoring for at least the last several hundred years, and we are now out of balance with our environment. With this realization, we can develop a plan, a vision of a future where humankind is back in balance with other species. "Let us put our minds together and see what life we will make for our children." This is a famous quote from Tatanka Iyotanka (Sitting Bull) in 1877, and it is valid for any century. We just have to start – now! What would your life look like if it were 100% sustainable? That is your goal, your vision for the future. Now figure out what steps need to be taken to get you there. What is the first step? Something you can do *today*?

Sustainability models

There are very few sustainability models intended for individuals to follow. There are many sites on the Internet that describe stuff to do, but no model to follow. Luckily, there are businesses that have "stepped up to the plate" and are trying to become sustainable. Either they made this decision because they really wished to be a source of inspiration for others, or they were forced to do so because of some environmental disaster and they had no choice—the whole world was watching. But either way, they are now striving to be on a sustainable path. They have established a vision and plan for the future. These plans follow sustainability models. Edwards (2005; 2010) describes many models that businesses, governments, and communities, etc., are using.

I highly recommend that you develop an individual sustainability model for yourself and your family. There is a difference between developing a model and implementing a model. A model shows you what has to be accomplished. Implementation tells how you are going to do it. For example, a model may suggest that you reduce wastes. Implementation asks you to consider each purchase and the packaging wastes that are generated from that choice. The best choice for obtaining green beans is to grow them yourself or buy them at a local farmer's market, versus purchasing them at the grocery store where they are packaged in a can, a plastic bag, or a waxed paper-lined cardboard box. Reducing wastes means that you take all of these choices into consideration before you decide.

For business models, there are so many to choose from! All these models start by asking similar initial questions. Doppelt (2003) lists these questions as:

- How sustainable are we now?
- How sustainable do we want to be in the future?

- How do we get there?
- How will we measure progress?

For individual models, there are none to choose from, but the same questions apply! Our initial questions should be the same:

- How sustainable are we now?
- How sustainable do we want to be in the future?
- How do we get there?
- How can we tell if we are making progress?

As individuals, we are motivated by the desire to have well-being for our families. We want our towns and neighborhoods to have economic vitality. Because what individuals want is linked to the vitality of the town, this frequently attracts the attention of the local leaders (Edwards, 2010). If you care, it spreads!

Some companies are leading the way, including Interface Inc. A 7-part strategy guides them towards their goal of becoming the world's first truly sustainable company. The 7 faces of "Mount Sustainability" from Anderson (1994) are as follows:

1. *Achieve* zero waste.
2. *Generate* benign emissions.
3. *Shift* to renewable energy.
4. *Close the loop* (closed-loop recycling).
5. *Use* resource-efficient transportation.
6. *Expand* the sensitivity hookup (community service).
7. *Redesign* commerce (lease).

If the individual wants to use a business model as a potential format for an individual model, only minor modifications to the principles are needed to address individual issues instead of company-wide concerns. The individual has to determine where wastes are generated. He or she has to consider where emissions are generated and find a way to make them benign, etc. The first 6 principles can be addressed individually in a fairly straightforward manner. However, the final principle needs extra thought. From an individual viewpoint it means that the way commerce is engaged is redesigned: if a product can be leased instead of purchased, the manufacturing company retains responsibility for the product beyond manufacture to use through disposal. This method helps reduce your own waste stream. (This idea will be addressed in more detail when we look at the things we make and purchase). But before it can be accomplished by the individual, we need a visible green grading system. "Green labeling" would inform us as to the responsible choices various manufacturers are making to create a structure within which the individual can also make good choices. Right now, that potential system does not exist or is fragmentary at best.

We have been led to believe that in order to "save the world" we will have to give up much comfort and many jobs, when in reality just the opposite is true. When you actually look at the numbers, it turns out that good environmental citizenship is great for the bottomline (Doppelt, 2003). If a company is going to stand out now, the path is green. Perhaps, if the individual is going to stand out, the path is also green.

Can we look to businesses in order to avoid the mistakes they have made? Is making money the only motivation that we have as individuals? We blame businesses for caring only about the bottomline. Unless our core values nourish and protect ALL people as well as the planet, we will struggle to get on a sustainable path (Doppelt, 2003).

Businesses are being challenged to re-evaluate how they operate and measure success. Their focus is shifting from environmental compliance to ecological, economic, and social responsibility. The demand for accountability is clearly a main driver for sustainable business practices. Now society demands transparency as well (Edwards, 2010). As an individual, can we step up to these same goals?

On December 18, 2010, a wave of revolutionary demonstrations, protests, and riots began in the Arab World, called Arab Spring (Figure 2.1). On September 17, 2011, the Occupy Wall Street movement began on Wall Street, and spread to over 95 cities, 600 communities and 82 countries (Adam, 2011). Signs saying, "Sorry for the inconvenience. We are trying to change the world" were displayed around the world. Those efforts can still be followed at http://www.occupytogether.org/.

It is clear that, worldwide, we are ready for change. According to James (2004), in order for change to occur, every community needs leaders to step up to inspire others. These special citizens have a passion for sustainability and are willing to work hard to make change happen in their communities. They are "fire souls," and every community needs at least four or five of them (James, 2004). Communities need to protect these individuals, or they can burn out. When an individual becomes inspired, they have a responsibility to pass that "fire" onto others so that not just one person has the responsibility to lead change. Leaders should run a "relay race," passing responsibility to others, and re-entering the race as needed and personal energy permits.

Albert Einstein said, "If we are to solve the problems that plague us, our thinking must evolve beyond the level we were using when we created those problems in the first place." In other words, we have to be smarter than the scientists of the last 200 years. That may not be as hard as it sounds. Sometimes it is something quite simple. One factory labeled the light switches so that everyone could see which switches controlled which lights. They saved $30,000 the first year (Hawken et al., 1999). We just need to rethink the way we do everything!

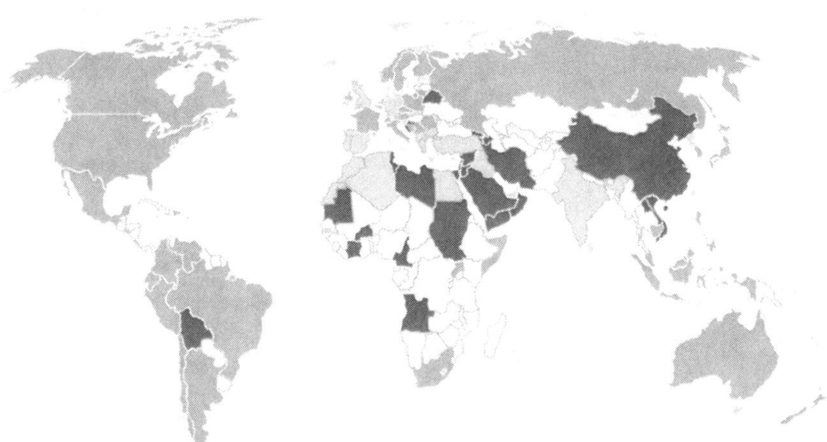

Figure 2.1. The Arab Spring and Occupy Movement impact worldwide: Black: Arab Spring; Lightest gray: Reaction to Arab Spring; Medium gray: Occupy Wall Street Movement. Source: Locations of Global 2011 Protests. Creative Commons Attribution-Share Alike 3.0 Unported license.

There are many models driving businesses toward sustainability, for example:

- The Precautionary Principle,
- The Natural Step's Four System Conditions,
- The Houston Principles, and
- The Coalition for Environmentally Responsible Economies Principles (CERES).

How can all this help us to develop an individual sustainability plan? We can use the business and community models to guide us in the development of our own sustainability model. Many are provided in summary by Edwards (2005; 2010). Look at each principle closely and translate the principle into a form that makes sense at the individual level. No existing model is complete. You may have to combine several plans (models). This is just a starting point. Remember that there is a difference between developing a model and implementing it. Develop the model so that it can be followed by anyone. We are each at different levels of sustainability. Some are just starting; others have been developing this new lifestyle for some time. A model, made up of several principles, should help anyone, including yourself, at any level of understanding. One principle within the model will mean one thing at one stage of understanding and will mean something else years later.

For example, the concept of "zero waste" will have one meaning for a company that has manufacturing as its business and another meaning for the individual. Zero waste for a company provides a new way to look at all incoming materials and the resulting waste stream. Anything discarded is a wasted resource, and a lost financial opportunity. Zero waste at the individual level means you work towards having no trash to take out each week. What is in your trash? How can you direct your purchases in the future so that you no longer have that waste? To implement the principle, you will need to change your buying habits. Your model will still have the same principle as the company model: Zero waste. Later, zero waste may mean no wasted energy. Or no wasted water. No wasted time. No wasted opportunities to be kind to another human being. No wasted effort to protect other creatures at risk. Our current level of consciousness will determine what zero waste means.

References Cited for Chapter 2
Importance of Vision and Sustainability Models

Adam, Karla. (October 16, 2011). Occupy Wall Street protests continue worldwide, Washington Post. Retrieved from http://articles.washingtonpost.com/2011-10-16/world/35278506_1_protesters-violent-clashes-worldwide-demonstration

Anderson, Ray. (1994). Interface story. Retrieved from http://www.interfaceglobal.com/sustainability.aspx

(The) Automatic Earth. (2012). The Occupy Movement of Mutual Knowledge. Retrieved from http://www.theautomaticearth.com/Psychology/occupy-movements-of-mutual-knowledge.html

Barker, Joel A. (2013). The Power of Vision. Retrieved from http://www.vision-album.com/

Burke, Julian. (2009). Success Secrets – The Importance of Having a Vision. *Visions.* Retrieved from http://www.dreammanifesto.com/success-secrets-importance-vision.html

Doppelt, Bob. (2003). *Leading Change Toward Sustainability: A Change-Management Guide for Business, Government and Civil Society.* Sheffield, UK: Greenleaf Publishing Limited.

Edwards, Andres. (2005). *The Sustainability Revolution: Portrait of a Paradigm Shift.* Gabriola Island, BC, Canada: New Society Publishers.

Edwards, Andres. (2010). *Thriving Beyond Sustainability: Pathways to a Resilient Society.* Gabriola Island, BC, Canada: New Society Publishers.

Evolution Training, Peak Perforance Through People Training. (2013). Retrieved from http://www.e-volution-training.co.uk/the-importance-of-vision-c55.html

Hawken, Paul, Amory Lovins, and L. Hunter Lovins. (1999). *Natural Capitalism*. New York, NY: Little Brown and Company.

Interface. (2010). *Mission Zero–The Power of a Challenging Vision: The Endless Possibilities series.* Retrieved from http://www.interfacecutthefluff.com/wp-content/uploads/2010/05/MissionZero.pdf

James, Sarah and Torbjörn Lahti. (2004). *The Natural Step for Communities: How Cities and Towns can Change to Sustainable Practices*. Gabriola Island, BC, Canada: New Society Publishers.

Keller, Valerie. (2013). *Backcasting to the Future*. Huffington Post, World Economic Forum. Retreived from http://www.huffingtonpost.com /valerie-keller/backcasting-to-the-future_b_3015954.html

English dictionary (online). (2009–2013). Macmillan Publishers Limited. Retrieved from http://www.macmillandictionary.com/us/dictionary/american/vision

Natural Step. (n.d.). *The Science Behind Our Approach*. Retrieved from http://www.naturalstep.ca/the-science-behind-our-approach

Steffen, Alex. (2011). *World Changing: A User's Guide for the 21st Century*. Abrams, NY: Sagmeister Inc.

Weiner, Jeff. (2012). From Vision to Values: The Importance of Defining Your Core, LinkedIn. Retrieved from https://www.linkedin.com/today/post/article/20121029044359-22330283-to-manage-hyper-growth-get-your-launch-trajectory-right .

CHAPTER 3
NATURAL ECOSYSTEMS

How are ecosystems changing? Where is the evidence of their change? The most powerful example is caught in time-lapse photography of the glaciers melting around the world. A sampling can be viewed at http://extremeicesurvey.org/ or in the full 2-hour movie "Chasing Ice" (Orlowski, 2012). The Extreme Ice Survey (EIS) team installed 27 time-lapse cameras at 15 sites in Greenland, Iceland, Alaska, and the Rocky Mountains, and took photographs annually in Iceland, the Alps, and Bolivia (Balog, 2012). The stunning photography shows without doubt the changing landscape in our ice-covered environments. This may also be the first time that global climate change has been documented in such a clear manner. The beauty of the glaciers is being lost on a daily basis, as the summer melts are not rebuilt in winter (Figure 3.1). Balog describes the link of this melting to climate change in a YouTube video: *The North Face: Extreme Ice Survey, Columbia Glacier, Alaska,* http://www.youtube.com/watch?v=LIHELl448r4

The melting of the glaciers represents an entire biome that is being lost. But ecosystems are being lost at every scale. If we are to save and restore them, we need to understand the economics involved in all of the services that these systems provide for free.

Figure 3.1. The loss of ice from the leading edge of a glacier is called glacier calving or iceberg calving. Source: Shutterstock.com.

What do ecosystems do for us? What are the services that nature provides to all organisms? We do not have to sign a contract with nature in order to obtain these services, although a hidden contract may be in place. For example, if we abuse our privilege to have access to these services, the service may be revoked!

Services provided through the normal functioning of our ecosystems are described in detail by the United Nations 2004 Millennium Ecosystem Assessment report that can be found online (www.millenniumassessment.org). The research in this 4-year study involved more than 1,300 scientists worldwide. They grouped ecosystem services into four broad categories:

1. *Provisioning* services provide food and water.
2. *Regulating* services control climate and diseases.
3. *Supporting* services are nutrient cycling and crop pollination.
4. *Cultural* systems provide spiritual and recreational benefits. (Environmental Mainstreaming, 2007)

In 1997, Costanza et al. did a crude calculation of the economic value of the services provided by 17 ecosystem services. Their valuation reflected astonishing numbers, in that the gross world product was about 18 trillion dollars and the services provided by ecosystems were valued at almost double that amount at 33 trillion dollars. To replace the functions provided free by nature would cost twice the total gross world product! The highest valued lands were the wetlands, valued at $7,924 per acre per year for flood control, storm protection, waste treatment, recycling and water storage (Hawken et al., 1999).

Now there are many methods used to estimate the value of the services ecosystems provide. According to the website "Ecosystem Valuation," (n.d.) the following are potential methods to use when attempting to understand the value of an ecosystem:

- *Market Price Method:* considers the products or services that an ecosystem provides for commercial markets.
- *Productivity Method:* considers the ecosystem products or services that are required in the production of commercially marketed goods.
- *Hedonic Pricing Method:* values services that an ecosystem provides that have a direct effect on the price of a product, i.e., how housing prices are affected by a local environment.
- *Travel Cost Method:* values ecosystems used for recreation.
- *Damage Cost Avoided, Replacement Cost, and Substitute Cost Methods*: costs avoided due to an ecosystem service; replacing or providing a substitute for an ecosystem service.
- *Contingent Valuation Method:* estimates the value of any ecosystem or service provided by that ecosystem based on willingness to pay for the service (the most widely used valuation method for "passive use").
- *Contingent Choice Method:* estimates the value of any ecosystem or service provided by that ecosystem by comparing tradeoffs within sets of services.

- *Benefit Transfer Method:* values for ecosystem estimates are determined by studies developed for another location.

Ecosystem Valuation (n.d.)

An ecosystem function currently appearing in the news on a regular basis is that of pollination. Ecosystems provide habitat for bees, the pollinators of flowering plants, which include most of our food crops. Agricultural crops have replaced many natural ecosystems. According to the United Stated Department of Agriculture (USDA, 2013) beekeepers who transport their bee colonies around the country to provide pollination services have been reporting hive losses of between 30% and 90%. This is called Colony Collapse Disorder (CCD). Greater than 50% of all affected colonies show symptoms inconsistent with any known cause of honey bee death. The bees just disappear. In some countries the collapse is close to 80% of the colonies. And yet in other countries, such as Canada and Australia, there is no CCD present. Of the 48 lower United States, 29 have CCD. Many theories exist about the cause of the disorder. According to the USDA (2013), the main directions for scientific study include:

- *Pathogens:* a higher total pathogen load of viruses and bacteria correlates with CCD more than any specific pathogen;

- *Parasites:* Varroa mites or the viruses that Varroa mites transmit may be a factor in causing CCD;

- *Management stressors:* bees have poor nutrition due to apiary overcrowding and stress brought on by transporting the bees to multiple locations across the country; and

- *Environmental stressors:* pollen/nectar scarcity, lack of diversity in pollen/nectar, availability of only pollen/

> nectar with low nutritional value, and limited access to water, or access to only contaminated water. Stressors also include accidental or intentional exposure to pesticides at lethal or sub-lethal levels.
>
> USDA (2013)

Since 2007, CCD has killed more than 10 million beehives in North America (Nesbit, 2013). In a University of Maryland study, pollen was collected from 7 sites where CCD is particularly severe. The pollen, which was then fed to healthy bees, "contained an average of nine different types of pesticides and fungicides. One pollen sample had 21 different chemicals" (Nesbit, 2013). The researchers discovered that bees that ate the fungicides, (labeled as harmless to bees) were 3 times more likely to be infected with the parasite known to cause Colony Collapse Disorder (Nesbit, 2013).

Honey bees are not native to the New World; they came from Europe with the first settlers. They are more prolific than the native pollinators and easier to manage on a commercial level for pollination of a wide variety of crops (USDA, 2013). As the bee population collapses, the crops that bees pollinate are also at risk (Figure 3.2).

What is the best action that the public can take? Research what plants provide good sources of pollen in your ecoregion. Pollinator Partnership (1996-2013) provides free booklets to each of 64 ecoregions in the United States describing trees, shrubs, vines, perennials, and crops that provide nectar and pollen for bees. Good sources include red clover, foxglove, bee balm, joe-pye weed, and other native plants (USDA, 2013). Of course, adding a beehive to your property increases their chances of survival as well, especially if the bees are exposed to organic crops rather than those laden with chemicals.

Fruits and Nuts	Vegetables	Field Crops
Almonds	Asparagus	Alfalfa Hay
Apples	Broccoli	Alfalfa Seed
Apricots	Carrots	Cotton Lint
Avocados	Cauliflower	Cotton Seed
Blueberries	Celery	Legume Seed
Boysenberries	Cucumbers	Peanuts
Cherries	Cantaloupe	Rapeseed
Citrus	Honeydew	Soybeans
Cranberries	Onions	Sugar Beets
Grapes	Pumpkins	Sunflowers
Kiwifruit	Squash	
Loganberries	Watermelons	
Macadamia nuts		
Nectarines		
Olives		
Peaches		
Pears		
Plums/Prunes		
Raspberries		
Strawberries		

Figure 3.2. Agricultural crops that are pollinated by the honey bee. Source: http://www.nrdc.org/wildlife/animals/bees.asp

A question that has not been asked in relationship to CCD is: if these chemicals are killing bees, what are the chemicals doing to us? How is our immune system affected as well? If we cannot wash off the chemicals before we eat the fruit or vegetable (Schoker, 2011), then we also are at risk in ways we may not yet understand.

Think about it for a minute. Where are the native ecosystems? We have replaced the prairies with corn and wheat. We clearcut our forests and replant them with a monoculture of trees. We have caught most of the fish in the oceans. The fertilizer and pesticide run-off from our fields are polluting streams and rivers. We have filled in 80% to 90% of the wetlands. The extra carbon dioxide in the air is acidifying the ocean ecosystems. We are at a time when we have to begin to work very hard to restore health to these environments, and reestablish a new relationship with nature.

How does a natural ecosystem work? What are the components? The soil holds a mixture of plants, some of which are nitrogen-fixers. Some gather micro and macro nutrients, the accumulators, and others are mulch plants. There are species from other kingdoms present, including insects that attract other wildlife; there are seed and forage species. The ecosystems harvest water. The natural system captures, cleans, and stores water for future use. This ecological garden captures nutrients and holds them in the soil. There are many layers from ground covers to herbaceous plants, ferns, shrubs, small trees, and canopy trees. Then you add all sorts of animals, and create niches specific to their needs (Hemenway, 2009). If we want to restore the earth to health, we have to make the ecosystems what they once were. We have to learn to garden the way nature gardens.

Take the country of Ecuador. Do you believe that we (in the United States) are more environmentally advanced, or is Ecuador more advanced? In the 1960s and 1970s, Texaco began exploring for oil in northeast Ecuador, in an area that was inhabited only by indigenous people. Toxic waste pits and polluted water sources were left in the wake of Texaco's abuse to that country. In 2008, Texaco was accused of not only widespread pollution, but deforestation and cultural destruction as well. The damages increased to 27 billion dollars (Gullo, 2008). As a result, the voters in Ecuador ratified the world's first eco-constitution, and Ecuador's constitution now recognizes that "ecosystems possess the inalienable and fundamental right to exist and flourish, and that people possess the legal authority to enforce those rights on behalf of ecosystems" (Community Environmental Legal Defense Fund, 2008).

We are trying to restore ecosystems in the United States too, and sometimes in the most unlikely of places. This field of endeavor, called restoration ecology, is happening in the suburbs of Chicago where a 30-year plan hopes to convert some of the area back to oak savanna. A total of 17,000 acres have been replanted with an ultimate goal of 100,000 acres (Chicago Wilderness, 2013).

Some other areas may not be as successful. Puget Sound in Washington State was just a couple of decades ago embarrassingly bountiful. Now it is at an ecological tipping point, on the brink of collapse due to overfishing, toxic chemicals, and invasive species (Steffan, 2011). The key to saving this environment is to save the shorelines that represent an important habitat for fish and birds. Along a 54-mile stretch of shoreline, 4 miles have been acquired for parks and shoreline restoration (Steffan, 2011). This is a great start. However, Puget Sound has 2,500 miles of shoreline. Developing greater awareness of the environment is paramount if greater success is the goal.

There are many critical environments that need our attention. Some of these are called biodiversity hotspots. This concept was introduced in 1980. Biodiversity hotspots hold especially high numbers of endemic species (species unique to that location) (Edwards, 2010). They originally covered 16% of the Earth's land but their remaining habitat now has been reduced to only 2.3%. Each location is severely threatened and has lost at least 70% of its native vegetation (Conservation International, 2013). Over 50% of the world's plant species and 42% of all terrestrial vertebrate species are endemic to the 34 biodiversity hotspots (Edwards, 2010). To protect just these 34 locations would require 34 billion dollars per year (Conservation International, 2013).

When we think about protecting biodiversity, we naturally think of the tropical rainforests in Brazil. We do not have to go to Brazil to find the endangered species. They are in our own backyards. Endangered Species International (2011) lists all of the animal species that are in danger as well as those that are on the extinct list. A serious wake-up call of greater interest on the Endangered Species website are those animals that are newly extinct (since 1500 AD). The website has an interactive map that leads to the lists of endangered species in every state (Kupris, 2002). Earth's Endangered Creatures (EEC) lists all of the species by country that are on the endangered list (EEC, 2014). This list seems to be more current, showing 714 plants and 1,555 animals in the United States.

In Oregon we have 37 animals and 14 plants on the endangered species list (Oregon Species, 2014). We truly only have to go outside our door to find a habitat that needs protection. Every year we lose about 2 million acres of land to highway construction, new shopping centers, and housing. Farms, forests, wetlands, and open spaces are destroyed to increase urban expansion (Edwards, 2010). Some small land trusts (both private and non-profit) are taking responsibility to set up permanent management for protected lands (Edwards, 2010). Those small non-profit volunteers that frequent our doorsteps, asking for a small donation, may be the only ones paying real attention to a needed action and perhaps deserve an extra minute of our time or funds.

Forests need protection everywhere, especially old-growth forests. Besides the full range of ecosystem services that they provide, they also sequester enough carbon from the air to offset 25% of worldwide CO_2 emissions. These natural services are many times more valuable than the value of wood fiber, especially when that fiber becomes a throwaway wrapper from a hamburger or an envelope from an unwanted credit-card solicitation (Hawken, 1999). How should we put a value on nature? There are three ways to look at the question (Edwards, 2010):

1. As a natural resource: what can we take from the land?
2. As real estate: what can we build on the land?
3. For ecosystem services: what can the land give us if it is restored to health and is protected?

There are some services for which there are no substitutes at any price. We have just discussed the service of the pollinators. Another is oxygen production by green plants. When Biosphere 2 was built in Arizona, we discovered just how incapable we are at creating life-supporting systems. It was originally built to be an artificial, closed ecological system to study space colonization and the manipulation of ecosystems, without harming the Earth. It operated from September, 1991 to September, 1993 and it failed miserably. The best scientific minds in the world could not construct a functioning ecosystem that could keep 8 people alive for 24 months. We add 8 people to the planet every 3 seconds (Hawken et al., 1999). We have no idea how to manufacture a watershed or wetland, topsoil, riverine systems, or pollinators, let alone a whole ecosystem (Hawken et al., 1999). Natural systems are extremely complex.

For example, a teaspoon of good grassland soil may contain 5 billion bacteria, 20 million fungi, and 1 million protists. If you expand that teaspoon to a square meter, the soil now contains perhaps 1,000 each of ants, spiders, woodlice, beetles and their larvae, and fly larvae; 2,000 each of earthworms, millipedes and centipedes. It will also contain 8,000 slugs and snails; 20,000 pot worms; 40,000 springtails; 120,000 mites; and 12 million nematodes (Eisenberg,

1998). Although we cannot create an ecosystem, we can create the environment necessary in order to give nature a chance.

When we built all the major dams in this country, we knew that, in approximately 100 years, the silt would accumulate behind the dam and significantly reduce the capacity of the dam to function correctly. Those 100 years are up, and that is why we hear so much about taking down the dams now. The smaller dams are easier to consider for removal. When the Marmot Dam was removed from the Sandy River in Oregon, scientists were impressed with how quickly the river washed the sediment away and how quickly the river looked and behaved naturally again (Millstein, 2008). It was thought that 2 to 5 years would be needed to absorb all the sediment, but the river did it months instead, with no harm to life in the river downstream. Salmon were seen swimming upstream the very next day after the dam removal. This is called restoration ecology. All we have to do is just give nature a chance!

In many parts of the world, clearcutting of forests takes place. It is well documented that this practice is detrimental to habitats, and yet clearcutting continues. Botkin and Keller (1997) describe in detail the benefits and liabilities. "Clearcutting and other even-aged systems [trees all planted at the same time] may have greater impacts on soils, water, and aesthetics, and result in different plant and animal communities than do selection harvesting systems" (Botkin and Keller, 1997). They begin their research survey by explaining that clearcutting has been controversial since the 1960s. One would think we would have made some progress in resolving the issues within the past 50 years!

Figure 3.3. Marmot Dam (upper left) and its removal from the Sandy River in 2008 (upper right).Sandy River today (lower left). Source: (u.l.) D Tanner, USGS; (u.r.) USGS; (l.l.) Rick Swart, ODFW.

There are some forests that are handled differently. In Wisconsin, the Menominee Tribe has practiced sustained yield forestry for more than 140 years on their 235,000 acre reservation. Their goal is to maximize both the quantity and the quality of their harvests under sustained yield management principles, while maintaining the diversity of native species. After years of logging, 2 billion board feet of timber have been produced, and yet the forest tree quality and the species diversity continue to improve. In the forest itself, 1.5 billion board feet of timber are still standing. David Grignon, a tribal planner, says, "The forest today is what it was 200 years ago when the Old Ones looked at it. Sometimes I go to just sit there and look around and I know that" (Marshall, 1993).

In aquatic ecosystems, the same concerns apply. A billion people (one-sixth of the world’s population) depend upon fish as their main source of protein (Edwards, 2005). Edwards in 2005 reported that in the last 50 years, 90% of all large ocean predators, tuna, marlin, swordfish, sharks, cod, halibut, skates, and flounder, had been over-fished by industrial fleets. That was devastating news. However, in a 1996 law, the Magnuson-Stevens Fishery Conservation and Management Act, a 10-year goal was set to rebuild each species under threat of extinction. The law ordered limits on catches until the fish populations had rebounded.

The benefits of that moratorium have been shown in the recent report from the Natural Resources Defense Council. They concluded that almost half of the species under watch (21 of 44 species) have met the targets set for population rebuilding, and 7 more species have increased their populations by at least 25%, a significant improvement (Wines, 2013). Other species have made very little progress, including 10 species off the New England coast. Although their numbers have increased more than 25%, Atlantic fish, including cod and flounder, are still being overfished (Wines, 2013).

What is inspiring about both the Menominee forest and the return of large predator fish is that we DO have other options, and they work! In every case, when we really search to find the correct, sustainable path, we find that it is totally possible. So we need to think about all views. We aren't trying to save the salmon in Oregon just so that we can eat them all, are we? We want to save them so that they can return to the numbers they once were.

We need to begin to use nature as a guide. In mature ecosystems, cooperation is just as important as competition (Benyus, 1997). Organisms develop unique niches so that everything is used and there is no waste in any system. Should one organism fail in a niche of the ecosystem, there is at least one other organism already in place to succeed it. Benyus (1997) developed a sustainability model based on nature called "biomimicry." If we follow the biomimicry principles, we can come more into balance with our environment. The principles as described by Benyus are as follows:

- Nature runs on sunlight;
- Nature uses only the energy it needs;
- Nature fits form to function;
- Nature recycles everything;
- Nature rewards cooperation;
- Nature banks on biodiversity;
- Nature demands local expertise;
- Nature curbs excesses from within; and
- Nature taps the power of limits (Benyus, 1997).

We can, even at the smallest of scales, begin to live by these principles. Take, for example, "Nature runs on sunlight." We have the opportunity to switch to solar power (by installing panels on our houses) or to wind power, made by the sun heating air and causing it to rise. In some states, wind power is an option through "the grid" and all that is required is to sign up for it. So the sun can power our lives. We can learn to recycle everything. To start, we can use banana peels and discarded newspapers to make new soil from composting. We can create diversity in our environment by reducing the amount of grass in our yards. By adding plants to our balconies, neighborhoods, business parks, and towns, we can increase and improve wildlife habitat. The National Wildlife Federation actually has a certification process whereby you can have any space, even your balcony designated as a "certified wildlife habitat" (NWF, 1996-2013). All that is required is to:

1. Provide food for wildlife
    a. Native plants, seeds, fruits, nuts, berries, nectar (flowers);

2. Supply water for wildlife
    a. Birdbath, pond, water garden, stream;
3. Create cover for wildlife
    a. Thicket, rockpile, birdhouse;
4. Give wildlife a place to raise their young
    a. Dense shrubs, vegetation, nesting box, pond.

Once you have provided those elements to make a healthy and safe wildlife habitat, you can earn the distinction of being part of NWF's Certified Wildlife Habitat® by completing their certification form online. There are a variety of signs available to inform others that your "space" has been certified (NWF, 1996-2013).

Another way to increase planted areas can occur the next time your roof needs to be replaced: replace it with an ecoroof. The Portland Ecoroof Program describes its benefits as: "An ecoroof significantly decreases stormwater runoff, saves energy, reduces pollution and erosion, and helps preserve fish habitat. Ecoroofs also absorb carbon dioxide, cool urban heat islands, and filter air pollutants. They increase habitat for birds and insects and provide much-needed greenspace for urban dwellers" (City of Portland, 2013). A booklet detailing ecoroof construction and legal requirements is available from the City of Portland at http://www.portlandoregon.gov/bes/article/331490.

There are times when just sitting back and waiting for a new study to show that we need to change, or a new law that says we have to change, is not enough, nor moves us forward appropriately. We have to notice what needs to be changed, and become part of the change process ourselves. This may include becoming an activist. The Occupy Movement was a clear statement that many of us are ready for change. There are many books to guide this endeavor (Boyd, 2012; Sen, 2003; Van Derzee, 2010) and the Ruckus Society has many manuals that can be downloaded (Ruckus Society, 2003). All describe non-violent methods for change.

In the Northwest, people are fighting to keep coal trains out of these states. Coal that should remain in the ground (see Chapter #9 on Energy) is being mined in Montana and loaded into train cars; its ultimate destination being China where it can be burned to produce energy. If allowed through our state, railroad cars, losing an average of 225 pounds of coal from each car during a 567-mile trip, would contribute to our pollution problems. For a 135-car train, that's about 15 tons or 53 lbs per mile per car, 32 trains per day. The health effects of all that coal dust both to residents as well as the natural environment would be staggering. We may be required to physically protest to keep this from taking place. Sometimes all we need to do is sign petitions and write letters. Either way, it is a form of protest that companies can no longer ignore. The speed with which news flies around the world because of the Internet has dramatically changed the response time of many companies.

There are also times when event participation is quite enjoyable. Every year at Christmas, there is the annual bird count across the nation. It has been a tradition for over a century. The 114th Christmas Bird Count took place from December 14, 2013, through January 5, 2014. During those 3 weeks, tens of thousands of volunteers throughout North America participate in this holiday tradition with the desire to make a difference (Audubon, 2013). The data gathered by the National Audubon Society allows scientists to assess migration patterns, locations, and populations. Eartheasy (2012) provides clear directions and materials necessary for anyone who wishes to participate. In eBird, a real-time, online checklist program, more than 40,000 individual birders across North America contribute 2,000,000 sightings per month (eBird, n.d.). In 2009, an iPhone application called BirdsEye was introduced to help to facilitate the count. Both of these programs were developed in part by the Audubon Society and the Cornell Lab of Ornithology.

The earth needs natural areas so desperately that any land we abandon quickly reverts back to nature. Some of the most remarkable transformations are those of involuntary parks. An involuntary park is a term coined by science fiction author and environmentalist Bruce Sterling to describe a land previously inhabited, but, due to environmental, economic, or political reasons, has been allowed to return to its native state. The demilitarized zone between North and South Korea is a good example. It has lacked any human habitation or activity for 60 years (Steffan, 2011). The complete removal of human activities has permitted a staggering array of rare plants and animals, including the highly endangered red-crowned crane, to proliferate (Steffan, 2011).

Rocky Flats Plant, a former nuclear weapons plant near Denver was shut down in 1989 due to environmental violations. Today the Rocky Flats site contains perhaps the largest remaining tallgrass prairie in North America, as well as several endangered or threatened species (Steffan, 2011). The nuclear wasteland of Chernobyl in the Ukraine is turning into an immense wildlife preserve. Many animals such as moose, wolves, and deer now roam in places where humans cannot safely walk due to the residual radiation, and more than 200 bird species (31 of them endangered) call the park home. Yes, there are significant mutations in the area. This involuntary park will not be safe for human habitation for 20,000 years. The flora and fauna that remain in the area (19 mile radius) will have time to evolve under the new conditions.

It is very important that we all increase the amount of natural area in our yards, communities, states, etc.. A step in the right direction is on Long Island, New York, where they have passed an ordinance that requires that 80% of each lot remain natural, and no more than 15% be planted as fertilized lawn (Roseland, 2005). The town of Seaside, Florida, was designed with no lawns allowed, only native vegetation and "public grass," grass planted in public areas for the enjoyment of all.

What might be a reason to "go native" on your property? You would save money on water, waste disposal, and various chemicals. You would save time because working with nature is easier and less work than struggling to maintain an area that works counter to nature. You would protect your health by reducing contact with chemicals. You would protect the environment, and in the process you would conserve water for salmon and other fish by keeping streams and lakes free of chemicals. Yard trimmings would be recycled into free fertilizer, and finally, you would create habitat for other species. With so many benefits, why do we cling to our monoculture grass lawns with such determination?

There are subtle ways to create more natural environments. If giving up your grass is not possible for some reason, make impervious pavement pervious. A range of sustainable materials and techniques are used for permeable pavements that allow the movement of stormwater to run through the surface. Runoff is reduced, and pollutants are removed from the water. This allows rainwater to return to the water table naturally instead of capturing the water and moving it as quickly as possible into the sewer. We can construct bioswales in low areas to remove silt and pollution from surface water and allow stormwater to infiltrate slowly into the ground. We can replace the grass strip between the curb and sidewalk with native plants.

Are there areas in your community, no matter how small, that can be re-established as natural habitats or ecosystems? Or around your home? Can you give 20-25% back to nature? How might you do that?

1. Take out even some of the grass lawn, and plant something appropriate to your local area.
2. Install bioswales to manage stormwater runoff naturally.
3. Plant more trees/shrubs, especially native plants.
4. Consider an ecoroof to increase more planted surfaces, and to insulate your house and reduce heating and cooling.

Look at every corner of your environment with new eyes. Look around your home, your workplace, your school, and where you shop to find new places for more plants, and more natural ecosystems.

Figure 3.4. We need to avoid looking at every natural ecosystem as only what it can provide humans. A forest is much more than just "boardfeet" and "paperpulp." Source: Bigstock.com.

References Cited for Chapter 3
Natural Ecosystems

Audubon. (2013). Christmas Bird Count. Retrieved from http://birds.audubon.org/christmas-bird-count

Balog, James. (2012). Extreme Ice Survey. Retrieved from http://extremeicesurvey.org/

Benyus, Janine. (1997). *Biomimicry: Innovation Inspired by Nature*. New York, NY: HarperCollins Publishers Inc.

Botkin, Daniel and Edward Keller. (1992). *Environmental Science, Earth as a Living Planet*. Third Edition. John Wiley and Sons Inc. Retrieved from http://www.wiley.com/college/environet/NATIONAL.HTM

Boyd, Andrew with Dave Oswald Mitchell. (2012). *Beautiful Trouble: A Toolbox for Revolution*. New York, NY: O/R Books.

Ecosystem Valuation. (n.d.). Retrieved from http://www.ecosystemvaluation.org/1-02.htm and http://www.ecosystemvaluation.org/dollar_based.htm

Environmental Mainstreaming. (2007). Retrieved from http://www.environmental-mainstreaming.org/Environment%20Inside/ Chapter%204/chapter4-5.html

Chicago Wilderness. (2013). Oak Savanna Restoration. Retrieved from http://www.chicagowilderness.org/who-we-are/corporate-council/day-of-service/

(The) City of Portland. (2013). Environmental Services. Portland Ecoroof Program. Retrieved from http://www.portlandoregon.gov/bes/44422

(The) Community Environmental Legal Defense Fund. (2008). Ecuador Approves New Constitution: Voters Approve Rights of Nature. Retrieved from http://www.celdf.org/article.php?id=302

Conservation International. (2013). Biodiversity Hotspots. Retrieved from http://www.conservation.org/where/priority_areas/hotspots/Pages/hotspots_main.aspx

Costanza, Robert, Ralph d'Arge, Rudolf de Groot, Stephen Farberk, Monica Grasso, Bruce Hannon, . . . Paul Suttonkk & Marjan van den Belt. (1997 15 May). The value of the world's ecosystem services and natural capital. *Nature*, VOL 387.

Eartheasy: Solutions for Sustainable Living. (2012). Annual Bird Count. Retrieved from http://eartheasy.com/play_annual_birdcount.htm

Earth's Endangered Creatures. (2014). Retrieved from http://www.earthsendangered.com/search-regions3.asp?ID=US

e-Bird. (n.d.). Audubon and Cornell Lab of Ornithology. Retrieved from http://ebird.org/content/ebird/

Edwards, Andres. (2005). *The Sustainability Revolution: Portrait of a Paradigm Shift*. Gabriola Island, BC, Canada: New Society Publishers.

Edwards, Andres. (2010). *Thriving Beyond Sustainability: Pathways to a Resilient Society*. Gabriola Island, BC, Canada: New Society Publishers.

Eisenberg, Evan. (1998). *The Ecology of Eden*. New York, NY: Vintage.

Endangered Species International. (2011). Retrieved from http://www.endangeredspeciesinternational.org/overview4.html

Gullo, Karen. (2008, November 27). Chevron Estimate for Amazon Damages Rises by $11 Billion. Retrieved from http://www.bloomberg.com/apps/news?pid=newsarchive&sid=af8f.cmZHsHk&refer=us

Hawken, Paul, Amory Lovins, and L. Hunter Lovins. (1999). *Natural Capitalism*. New York, NY: Little Brown and Company.

Hemenway, Toby. (2009). *Gaia's Garden: Guide to Home-Scale Permaculture.* White River Junction, VT: Chelsea Green Publishing.

Kurpis, Lauren. (2002). Endangered Species. Retrieved from http://www.endangeredspecie.com/disclaimer.htm

Marshall, Larry Pecore. (1993, Spring). The Trees Will Last Forever: The integrity of their forest signifies the health of the Menominee people. *Cultural Survival Quarterly.* Issue: 17.1 Resource and Sanctuary. Retrieved from http://www.culturalsurvival.org/publications/cultural-survival-quarterly/united-states/trees-will-last-forever-integrity-their-fores#sthash.Z5JDOM7m.dpuf

Millstein, Michael. (2008, July 30). Oregon's Sandy River successfully reinvents itself after dam removal, Scientists are impressed how fast the river is digesting Marmot Dam sediment. Retrieved from http://www.oregonlive.com/news/oregonian/index.ssf?/base/news/1217390121239080.xml

National Wildlife Federation (NWF). (1996-2013). Wildlife Habitat Certification. Retrieved from http://www.nwf.org/certifiedwildlifehabitat

Nesbit, Jeff. (2013 August 7). Bee Colony Collapses Are More Complex Than We Thought, *U.S.News & World Report LP*. Retrieved from http://www.usnews.com/news/blogs/at-the-edge/2013/08/07/bee-colony-collapses-are-more-complex-than-we-thought

Oregon Species. (2014). U.S. Fish and Wildlife Service. Oregon Fish and Wildlife Office. Retrieved from http://www.fws.gov/oregonfwo/species/

Orlowski, Jeff (Director). Pesemen, Paula DuPre and Jerry Aronson, (Producers). (2012 November 16). *Chasing Ice* [Documentary]. United States: Submarine Deluxe in association with Diamond Docs

Pollinator Partnership. (1996-2013). Pollinator Friendly Planting Guides. Retrieved from http://www.pollinator.org/guides.htm

Roseland, M. (2005). *Toward Sustainable Communities: Resources for Citizens and Their Governments*. Gabriola Island, BC, Canada: New Society Publishers.

(The)Ruckus Society. (2003). *Action Planning Training Manual.* Oakland, California. Retrieved from http://www.ruckus.org/section.php?id=82

Schocker, Laura. (2011, June 13). Dirty Dozen: The 12 Fruits And Vegetables With The Most Pesticides. The Huffington Post. Retrieved from http://www.huffingtonpost.com/2011/06/13/dirty-dozen_n_875718.html#s290785title=Apples

Sen, Rinku. (2003). *Stir It Up: Lessons in Community Organizing and Advocacy*. San Francisco, CA: Jossey-Bass, A Wiley Imprint.

Steffen, Alex. (2011). *World Changing: A User's Guide for the 21st Century*. New York, NY: Sagmeister Inc.

USDA (2013). Honey Bees and Colony Collapse Disorder, Retrieved from http://www.ars.usda.gov/News/docs.htm?docid=15572

Van Derzee, Bibi. (2010). *The Protestor's Handbook*. London, England: Guardian Books.

Wines, Michael. (2013, March 15). Fish Populations in the United States Rebound. New York Times. Retrieved from http://www.nytimes.com/2013/03/16/science/earth/fish-populations-in-us-rebound-since-1996-catch-limits-law.html?_r=0

CHAPTER 4
WATER: CAPTURE, REUSE, AND PROPER DISPOSAL

What do you think are the sustainability issues regarding water? Where does your water come from? Do you use "city" water? Is your city water safe 100% of the time? Have you thought about capturing your own water? How many times do you use your water? What happens to your water when you are finished with it? Do you take it for granted that when you turn on the tap, clean, pure water will always be there? Are any of these questions something that you have not considered before?

There are many serious water issues around the world. Forty percent of people live where water is scarce, and 17% have no clean water (Edwards, 2010). More than half of the people in the world do not have the quality of water that Roman citizens had 2,000 years ago. One child dies every 15 seconds from a waterborne disease. Africans spend 40 billion hours every year collecting and hauling water. Water shortages will be felt by 66% of the world population by 2025 (Edwards, 2010) as the average supply of water per person drops by 30% (*Water*CAMP*us*, 2013).

While only 10% of our water supply is used for household consumption, 70% is used by agriculture (Edwards, 2010). Of the water used for agriculture, 93% is used to irrigate crops by flooding the fields, the least effective method of delivery (Hawken et al., 1999). If half of those acres were converted to drip or sprinkler irrigation, enough water could be saved to feed the extra 2.6 billion people we will have on the planet by 2025 (Hawken et al., 1999).

While all of the above conditions may be serious, they are not surprising based on our current population level. Our water quality is also declining as water treatment plants are not able to remove all contaminants, and are exposing us to an unwanted, vast array of pharmaceuticals (Donn, 2008). For example, in Portland, Oregon, 6 drugs were found, including estradiol, ethinyl estradiol (both used by women going through menopause), acetaminophen, caffeine, ibuprophen, and sulfamethoxazole (an antibiotic to treat infections and pneumonia)—all present in trace amounts in our water. The city of St. Louis recorded 11 drugs in their water supply; Denver had 12; Philadelphia recorded the most at 17 (Pharmaceuticals in Water, 2008). Many drugs are now prescribed in the form of creams that get washed into the water supply while showering, causing a much worse problem (Drugs, 2011). For example, one man using a testosterone cream can put as much of the hormone into the water as the natural excretions from 300 men (Drugs, 2011).

How bad is this problem? U.S. Geological Survey (n.d.) did a study in their Toxic Substances Hydrology Program (1999-2000). They took water samples from 139 streams in 30 states and found:

- Measurable amounts of one or more medications including antibiotics, antidepressants, blood thinners, heart medications (ACE inhibitors, calcium-channel blockers, digoxin), hormones (estrogen, progesterone, testosterone), and painkillers;

- Also found were caffeine (which, of course, comes from many other sources besides medications), carbamazepine, an antiseizure drug, fibrates to improve cholesterol levels, and some fragrance chemicals (galaxolide and tonalide) (Drugs, 2011).

There is also significant evidence that fish are being affected by these chemicals (Drugs, 2011). Is anything being done? The Natural Resources Defense Council has asked drug manufacturers to consider these issues and to design "eco-friendly" drugs that are better absorbed by the body and that will break down once they are excreted, therefore protecting the environment (Drugs, 2011).

Detailed information regarding water quality and quantity in the United States is not well known. The website *Water*CAMP*us* describes American issues and suggests a need for a "visionary new approach to management and conservation of water resources" (*Water*CAMP*us*, 2013). Their website, which shows a series of maps describing water concerns by county, is extremely informative and addresses the following issues:

- Combined Water Problems Map,
- Salting of Water Supply,
- Acid Rain,
- Aquifer Depletion,
- Arsenic Contamination,
- Elevated Nutrient Contamination,
- Energy Production Demand,
- Heavy Metals Contamination,
- High Demand,
- Mercury Contamination,
- Microbial Contamination,
- Mining Pollution,
- Organic and Industrial Toxins,
- Radioactive Contamination,
- Pesticide Contamination,
- Radon Contamination,

- Salt Water Intrusion into Aquifers (*Water*CAMP*us*, 2013).

The Ogallala Aquifer is a natural, underground water reservoir that extends from South Dakota to Texas. Forty percent of American beef cattle depend upon grain irrigated by the water from the Ogallala which is drawn down 3 to 10 feet per year due to this agricultural usage. The recharge rate is less than one half inch per year (Hawken et al., 1999). To put one pound of beef on the table, 100 pounds of topsoil is lost, plus more than 8,000 pounds of Ice Age-vintage groundwater. Consider how much beef you eat in a year. While some topics covered in these book chapters will show great improvement over the past 10-15 years, water is not one of those topics. In a recent publication, Konikow (2013) writes that water consumption has increased over the last decade. Konikow states "the rate of groundwater depletion has increased markedly since about 1950, with maximum rates occurring during the most recent period (2000–2008), when the depletion rate averaged almost 25 km^3 per year (compared to 9.2 km^3 per year averaged over the 1900–2008 time frame)" (Konikow, 2013). Clearly, our water consumption patterns are not sustainable, are getting worse, and deserve more attention than we are currently directing toward this issue. It is not an issue that can wait. The first rule in sustainability is conservation.

In some sections of Texas, water consumption was reduced by using a block of gypsum the size of a lump of sugar, buried near plant roots to monitor soil moisture content; water was delivered only when needed. A decade of using this method nearly halted the aquifer deletion in that area, saving 25% to 50% of their irrigation water (Hawken et al., 1999).

California has a tax on excess water use. Farmers bury a drip irrigation system to deliver one drop at a time, right where it is needed. This has allowed the state to sell back 32 billion gallons of water that were previously wasted each year (Hawken et al., 1999). In Oregon, the farmers save 10% to 20% of their water, even up to 40%, through education (a 3-hour consultation) and better management (Hawken et al., 1999). When New York City officials compared the costs of building a new water-filtration plant versus conserving the natural watershed, the watershed conservation was 10 times less expensive: 6 to 8 million dollars versus 660 million dollars (Edwards, 2010).

Water conservation and wastes are intimately linked. In Goleta, California, the citizens were required to build a new sewage treatment plant costing 1.5 million dollars. By implementing emergency drought measures, some technical improvements, and a little rationing, their water consumption was cut 30% in a single year, from 135 gallons per person per day to 90 gallons, twice their targeted goal (Hawken et al., 1999). Sewage flow fell 40%, and water savings eventually grew 40% (Hawken et al., 1999). It is amazing how we can conserve if we are told our property taxes are going to increase or that our rent is going to go up!

Businesses are beginning to work in the direction of conservation and restoration by changing their corporate turf lawns into grasslands with wildflowers that are drought tolerant (Hawken et al., 1999). The Coca-Cola Company reduced their water usage 79% in their Seattle facility by using air to clean the inside of the cans before filling them (Hawken et al., 1999). Gillette Company reduced the water needed to make razors by 96%, and to make Paper Mate® pens by 90%. A Kansas City steel mill changed their process so that they now use the water 16 times (purifying it in settling ponds between usages) saving 3.6 million gallons a day, even though their process uses 58 million gallons a day (Hawken et al., 1999).

At the individual level, xeriscaping is a method for designing low-water landscaping that is beautiful, provides natural cooling, fire protection, and native bird and wildlife habitat. The design does not have to be a cactus farm, commonly associated with xeriscape landscapes. Plants that are watered with efficient technologies or stored rainwater will also work to reduce water consumption (Hawken et al., 1999).

The average family of four in the U.S. uses 400 gallons of water per day, 70% of that indoors; that's about 70 gallons per person per day (Water Sense, 2008). That figure could be reduced to 52 gallons with minimal improvements; to 40 gallons with more efficient toilets, clothes washers, dishwashers, showerheads and bathroom faucets, plus a greywater toilet flushing system; and to 20 gallons per person if greywater were used to water plants outdoors (Hawken et al., 1999). The bathroom is the largest consumer of indoor water. New toilets use 60% less water than older models (Water Sense, 2008). A leaky toilet wastes 200 to 750 gallons a day, and a faucet, which often runs at 2 gallons per minute, can save 200 gallons per month if turned off while brushing your teeth (Hawken et al., 1999; Water Sense, 2008).

Water-saving washing machines save 50% more water than traditional models. Horizontal-axis washing machines use 40-75% less water (Hawken et al., 1999). If you wash the dishes by hand while the water is running, 20 gallons of water are used, whereas if you fill the sink or use a dishpan, you use 50% less water (Water

Sense, 2008). A leaky faucet uses an extra 300 gallons of water a month (Hawken et al., 1999).

However, the leaks in your house are nothing compared to those in the system that gets the water *to* your house. A good system loses 10% of its water en route; the average U.S. city loses 25% of its water in leaking pipes (Hawken et al., 1999). The amount in other countries is much worse, with Bombay losing one-third and Manila over half. In New York City, 57,000 miles of water mains were surveyed in the early 1990s. This resulted in the repair of 66 breaks, 671 leaks, and, as a consequence, saved New York 49 million gallons of water a day (Hawken et al., 1999).

Obviously, water conservation is critical, and an easy first step. There are three other important steps: capturing rainwater; building greywater systems for water reuse; and learning about proper disposal. In the future there is no such thing as wastewater. Rain falls naturally as distilled water. It is soft, pure, and no chemical treatment is needed to make it safe, unlike the water from our wells and reservoirs (Hawken et al., 1999). When it rains in our cities, we go to great expense to guide this water into gutters and sewers and big pipes where it combines with the waste products of humans and industry (Hawken et al., 1999). We have been told not to trust rainwater, and yet it has a hardness of zero, tastes fresh, and leaves faucets and tile sparkling clean. There are many companies developing rainwater harvesting at all scales, at the individual scale (Figure 4.1a) and the community level (Figure 4.1b and Figure 4.1c). Water harvesting can also be turned into art to water plants (Figure 4.1d), or to generate music. A building in Dresden, Germany, added

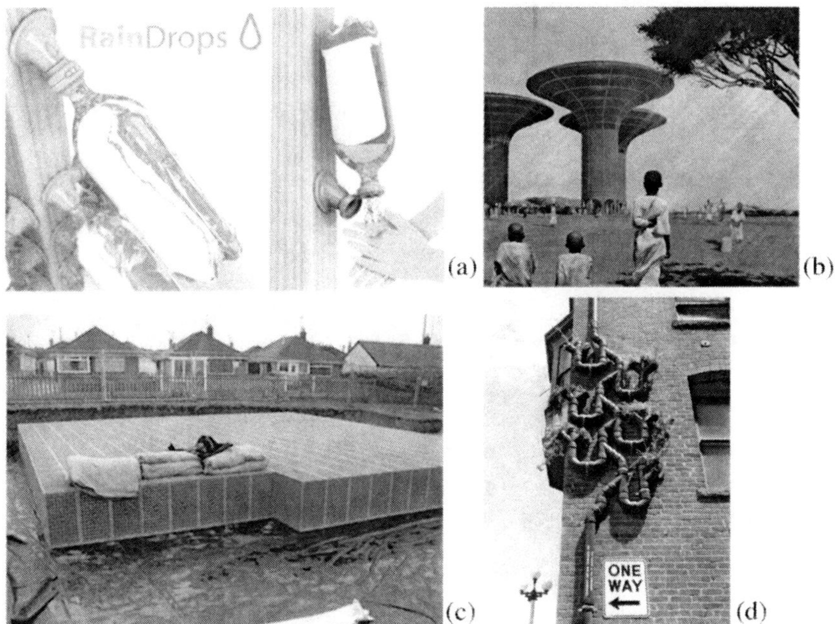

Figure 4.1. Capturing water. (a) RainDrops attach standard plastic bottles to existing gutter systems. The water flowing through the gutter is stored in the bottles to give you water for gardening and to wash your hands. Source: Evan Gant; (b) Water harvesting in the Sudan at the community scale. Source: Jordana, Sebastian, 2010. (c) Rainwater harvesting for community use. Source: Catalogue Engineering Ltd. (d) Downspout planter system. Source: Bigstock.com.

a beautiful rainwater sculpture to the façade of a building (Agovska, n.d.). If your city has rain, feature it!

Water collection apparatus can range from simple hanging bags to large above or underground storage tanks, or to wall containers that serve also to warm a patio when the weather turns cooler. If space is limited, there exists a water "pillow" that can slip under the house in the crawl space. In the Sudan, water collection is accomplished at the neighborhood scale with the huge collection towers (Figure 4.1b). When you capture your own water, you are in control of the quality.

How do you do the calculations to figure out how much water you can capture from your own roof? Hemenway (2009) provides directions where A = Area of roof (area of ground covered by the roof), and R = Annual rainfall (Figure 4.2).

	English measurements	Metric measurements
Area (A) covered by the roof	Width x length i.e. 30 ft x 36 ft = 1080 ft^2	Width x length i.e. 10 m x 12 m = 120 m^2
Annual rainfall (R)	Average in U.S. 35"	Average in U.S. .89 m
Potential volume of rainfall to be captured per year	A x R $\frac{1080\ ft^2 \times 35}{12} = 3{,}150\ ft^3$	A x R 120 m^2 x .89 m = 89.2 m^3
Convert to liquid volume	3,150 ft^3 x 7.5 = 23,625 gallons (gal)	89.2 m^3 x 1000 = 89,200 liters (L)
Emergency rationing for 1 person	23,625 gal ÷ 2 gal/day = 11,812 days	89,200 L ÷ 7.6 L/day = 11,737 days
Emergency water for 1 year	11,812 days ÷ 365 = 32 people	11,737 days ÷ 365 = 32 people

Figure 4.2. How to calculate the amount of rainfall you can capture from your roof. Source: Hemenway, 2009. Metric conversions through Kyle's Converter: online at http://www.kylesconverter.com/.

In an emergency situation, plan on 2 gallons of water per person per day (one to drink, the other for food preparation and cleaning/ bathing). Using the above example, the 23,625 gal (89,200 L) would provide enough water for 11,812 days (or 11,737 days for the metric figures). So the water that could be captured on a 1,080 ft^2 (120 m^2) home would provide the minimum water needed to keep 32 people alive for a year (Emergency Water Storage, 2013) (Figure 4.2).

One book that explains the details of rainwater harvesting in clear easy language is *Rainwater Collection for the Mechanically Challenged*, by Suzie Banks and Richard Heinichen (2004). Water conservation and capture are the easiest ways to reduce water usage. The next easiest way is to reuse our water as many times as we can before we discard it.

Mistakes we make with water are the same mistakes we make with energy. We are using a non-renewable resource as if it were unlimited. We need more water, so we look for MORE water; we don't do the obvious which is to use what we have more efficiently (Hawken et al., 1999). We use the best drinking water for every use, including flushing our toilets or washing our cars and driveways (Hawken et al., 1999).

Greywater is water from sinks, showers, and laundry, contaminated with a little soap, a few flakes of dead skin, and friendly bacteria that live mutualistically with us (Hemenway, 2009). Greywater can be saved and used again to water plants or flush toilets. The very simplest of systems is just a bucket. While waiting for the bathtub water to run hot from the faucet, I capture the cold water and then use it water to plants. In our house, that water amounted to a gallon every morning, and that gallon applied to the front lawn meant we had a green landscape when the rest of the neighborhood's landscape was brown. California passed a plumbing code to use greywater in 1994, cutting their total water use in half (Hawken et al., 1999). Oregon passed a similar law in 2011. We have yet to see the benefits from that law.

New greywater systems are in development. One example plumbs a bathroom sink with a holding tank that collects water to be used later to flush the toilet, making it possible to save up to 5,000 gallons per year (18,927 L) in a typical household (Bravenewleaf, 2008). There are many recirculating showers available that can lower not only how much water we use to shower, but also how much energy we use to heat that water (Polland, 2012).

Blackwater is water from toilets. Leaky toilets are found in 26% of households. A standard toilet uses 5 to 7 gallons (19 to 26 L) of water per flush, more water than many families around the world use for all their needs in an entire day (water-saver toilets use 1.6 gallons [6 L] or less per flush) (Hawken et al., 1999). There are several ways to have a functional, attractive toilet that uses no water. There are waterless urinals, separating toilets, composting toilets, and incinerating toilets. They can save 40,000 to 60,000 gallons (151,000 to 227,000 L) of water per unit per year (Hawken et al., 1999). With a composting toilet, the waste goes back to nature. With an incinerating toilet, very little waste is left, but nothing goes back to nature, and more energy is required during the incinerating process. Separating toilets prevent urine from mixing with solid matter. If we used these no-mix toilets, municipal wastewater treatment plants would have a much easier task: they could simultaneously produce methane to generate electricity, extract nutrients like phosphorus and nitrogen for use as fertilizer, and avoid using the fossil fuels currently required for fertilizer (Graham, 2010; Singh, 2012). When we look at energy in the future, we may see a system that uses human urine to fertilize commercially grown algae to generate biofuels.

We have looked at conserving water, capturing rainwater, reusing greywater, and reducing the amount of blackwater. Now we'll look at processing our own wastes. One way to do this is to build a backyard wetland. In nature, wetlands are used to clean and recycle water. The plants, animals, and microbes present purify the water while turning the pollutants into biomass. A family that takes several showers and does several loads of laundry per day may require about 200 gallons (757 L) of water per day (Hemenway, 2009). A 600-gallon (2,271 L) wetland (approximately 2 ft deep by 4 ft wide by 10 ft long (0.6 m deep by 1.2 m wide by 3 m long)) would serve this family's needs in waste processing. An article in *Inhabitat* (2012) describes 6 ways that human wastes can be converted into energy. Steinfeld and Del Porto (2007) describe more than 55 ways to process wastewater ecologically, from a single house that processes all its wastewater to cities that are powered by their wastewater. The idea is to begin to look at everything through new eyes. There is no waste in the future. Everything is reused and recaptured for future use, and water and environments are protected.

In Portland, Oregon, environments are protected when rainwater is diverted from the sewer into bio-retention swales, or bioswales. A bioswale is like a mini-wetland. More than 900 had been built in Portland by 2011 (WeBuildGreenCities, 2011). A wetland natural water filtration system can also eliminate the need for chlorine in swimming pools. The swimming pool has connected wetland pools, where plants are used to filter the water and keep it clean and clear (BioTop, n.d.).

Many problems result from anything that we direct to the sewer. Most cities' sewerage systems are hundreds of years old. These systems were designed before we understood the chemistry and microbiology involved in them (Hawken et al., 1999). In Los Angeles alone, over a billion gallons a day are collected through 6,500 miles of pipes. Treatment at the neighborhood scale or single-building scale might be a better idea. By switching from a chemical system to a biological one, clean wastewater could be provided at a lower cost while avoiding environmental safety hazards (Hawken et al., 1999).

When you think about a wastewater treatment plant, you probably do not picture a series of serene wetland planters in the lobby of an office building. The Tidal Flow Wetland Living Machine® is an example of wastewater treatment done exactly that way (Living Machine, 2012). Living Machine's website describes a tidal flow method and a hydroponic method of wastewater treatment. The Living Machine incorporates wetlands designed to promote micro-ecosystem development in an office building. The wetlands are constructed as part of an exterior landscape, part of an interior open space such as an office lobby, or as part of a greenhouse (Living Machine, 2012). The wetlands are alternately flooded and drained as water moves through the system. The micro-ecosystems remove nutrients and solids from the wastewater. The final stage involves filtration and disinfection to leave pure water ready for reuse. Sensors monitor water quality to ensure that reclaimed water is completely safe. Thousands of gallons (or liters) of water a day are processed, but the only things the casual observer sees are lush, vibrant plantings (Living Machine, 2012).

Biological treatment plants cost the same or less to build than traditional waste treatment plants, especially small systems (Hawken et al., 1999). Biotreatment plants yield fertilizers and soil amendments instead of hazardous toxic chemical wastes. Biological treatment plants also sequester heavy metals, and promote the growth of natural bacteria that kill pathogens. These facilities look like a water garden, wetland or greenhouse, and there is no odor (Hawken et al., 1999). The effluent (water leaving the biological system) yields a cleaner, safer, higher quality water. These methods lend themselves beautifully to commercial buildings, with gardens of cascading waterfalls, and tanks filled with fish, flowers and water plants (Hawken et al., 1999).

Our own city yards and lawns are excellent places to start conserving water. They are notorious water guzzlers, increasing our summer water bills substantially. Large reservoirs are required to supply this water to our homes. Pumps, hoses, and a homeowner's constant attention are also required. If we look at nature, at a forest or a prairie, how does it survive without all that attention? Where is water stored naturally? Lakes and ponds are the reservoirs. The soil is a natural, giant sponge for water, holding it for months to be used as needed by the environment. Plants absorb the water from the soil and transpire it into the air. The water is cleaned in the wetlands and, through evapotranspiration, is released back into the air by trees. When it rains, the water is again stored in the lakes, ponds, and soil. Obviously, the cheapest place to store water is in the soil (Hemenway, 2009), and key to this process is the organic matter (dried leaves and rotting wood). If the soil contains as little as 2% organic matter, irrigation can be reduced 75% (Hemenway, 2009).

You can mold your landscape to hold more water by contouring the land, and by adding shallow depressions (swales) to slow the water runoff and allow infiltration. The swales should be spaced closely together if you are in an area of high rainfall. For example, receiving 40 in. to 50 in. (100 cm to 127 cm) of rain per year would mean the swales should be about 18 in. apart (46 cm) (Hemenway, 2009). The swales can also be filled with straw or tree branches to absorb and retain more water in the environment.

Urban hydrology is beginning to appreciate the benefits of capturing rain as it falls, of returning rainwater to be stored as groundwater, and of greening a city. Water is no longer something to just get into a concrete pipe as quickly as possible. Water creates habitat. Water is life.

The evaporative power of water is being put to use in warm climates in order to refrigerate produce without electricity (Hawken et al., 1999). The Zeer Pot-in-Pot, originally used in Egypt around 3000 CE, is a small pot that is placed inside a larger pot. Sand and water fill the space between the two pots. A wet cloth placed over the top prevents excessive drying. As the water in the sand evaporates, the contents of the smaller pot (up to 26 lb or 12 kg of fruit and vegetables) are cooled (Figure 4.3). Austin Liu provides online directions to build a Zeer Pot (Liu, 2014).

In developing countries, having clean water solves several problems. Approximately 43% of the global population does not have access to clean drinking water (Vestergaard-Frandsen, n.d.). Vestergaard-Frandsen developed the LifeStraw®, a valuable tool for those who do not have access to clean drinking water. This YouTube describes their product and the additional benefits it provides: http://www.youtube.com/watch?feature=player_embedded&v=sH5SZViwTR0 Because the residents in these areas no longer have to boil their water, trees are being saved. The averted deforestation prevents additional global warming. Children are not exposed to as much carbon monoxide in the home during the boiling process (Vestergaard-Frandsen, n.d.). In addition, for those of us in the developed world, LifeStraws® provide safe drinking water while camping, hunting, hiking, or in emergency situations.

Figure 4.3. Zeer Pot-in-Pot is a non-electric means to refrigerate fresh fruits and vegetables. Source and instructions: http://bit.ly/zeerpot

Another beautiful water conservation video, this one filmed in India, shows that it is everyone's responsibility to address the issue of water capture. No matter how small the scale, we all can make a difference: http://www.youtube.com/watch?v=wWnhYIIKY0U

In the economically developed countries of the northern hemisphere (the Global North), most of the water we use is our "virtual" water, the water it takes to get stuff to us (Steffan, 2011). For example, if you think your morning coffee cup holds 12 oz (0.35 L) of water, think again. It has hidden water usage closer to 40 gallons (151 L) (Steffen, 2011). Where can you begin to reduce the amount of water you really use each day? Use a water footprint calculator to determine the hidden water in your lifestyle. Calculate the total amount of water you use at www.waterfootprint.org or http://environment.nationalgeographic.com

References Cited for Chapter 4
Water – capture, reuse, and proper disposal

Agovska, Dragana. (n.d.). Musical Façade "Gutter Funnel Wall, Dresden, Germany," in *Architecture, Art, Design*. Retrieved from http://www.architectureartdesigns.com/musical-facade-gutter-funnel-wall-dresden-germany/

Banks, Suzie and Richard Heinichen. (2004). *Rainwater Collection for the Mechanically Challenged.* Dripping Springs, TX: Town Publishing.

BioTop Natural Pools. (n.d.). Retrieved from http://www.biotop-natural-pool.com/referenz-teiche.html

Bravenewleaf. (2008). AQUS greywater kit makes heaps of sense, saves water & money. *Environment*. Retrieved from http://www.bravenewleaf.com/environment/2008/02/aqus-greywater.html

Donn, Jeff, Martha Mendoza and Justin Pritchard. (2008). PHARMAWATER I, Pharmaceuticals found in drinking water, affecting wildlife and maybe humans. Associated Press Writers. Retrieved from http://hosted.ap.org/specials/interactives/pharmawater_site/day1_01.html

Drugs in the water. (2011, June). *Harvard Health Newsletter*, Harvard University, Retrieved from http://www.health.harvard.edu/newsletters/Harvard_Health_Letter/2011/June/drugs-in-the-water

Edwards, Andres. (2010). *Thriving Beyond Sustainability: Pathways to a Resilient Society.* Gabriola Island, BC, Canada: New Society Publishers.

Emergency Water Storage. (2013). Homeland Security, Retrieved from http://www.nationalterroralert.com/safewater/

Hawken, Paul, Amory Lovins, and L. Hunter Lovins. (1999). *Natural Capitalism*. New York, NY: Little Brown and Company.
Hemenway, Toby. (2009). *Gaia's Garden: Guide to Home-Scale Permaculture.* White River Junction, VT: Chelsea Green Publishing.

Inhabitat. (2013). Six Ways to Convert Poo into Power, Retrieved from http://inhabitat.com/6-ways-to-convert-poo-into-power/no-mix-vacuum-toilet-e1340822738688/?extend=1

Jordana, Sebastian. (2010, 15 Mar). "Watertower / Hugon Kowalski" *ArchDaily.* Accessed 28 Apr 2014. http://www.archdaily.com/?p=52910

Konikow, Leonard F. (2013). Groundwater depletion in the United States (1900−2008): U.S. Geological Survey Scientific Investigations Report 2013−5079, 63 p., Retrieved from http://pubs.usgs.gov/sir/2013/5079/SIR2013-5079.pdf

Kyle's Converter. (2014). Retrieved from http://www.kylesconverter.com/

Liu, Austin. (2014). How to Make a Practical Zeer Pot, *Instructables*, Retrieved from http://bit.ly/zeerpot

Living Machine. (2012). Retrieved from http://www.livingmachines.com/About-Living-Machine.aspx

Pharmaceuticals in Water, 46 Million Americans Drinking. (2008). Retrieved from http://hosted.ap.org/specials/interactives/_national/pharmawater_update/index.html

Polland, Jennifer. (2012). This Futuristic Shower Will Save You Thousands Of Gallons Of Water. *Business Insider*. Retrieved from http://www.businessinsider.com/water-recycling-shower-uses-70-percent-less-energy-and-water-than-a-conventional-shower-2012-8#ixzz2cwl51yRY

Richard, Michael Graham. (2010). After Smart Grids, Smart Sewage? Urine-Separating NoMix Toilet Gets Thumbs-Up in 7 European Countries. *Treehugger*. Retrieved from http://www.treehugger.com /bathroom-design/after-smart-grids-smart-sewage-urine-separating-nomix-toilet-gets-thumbs-up-in-7-european-countries.html

Singh, Timon. (2012). Poo Power: New Toilet System Turns Human Waste Into Electricity And Reduces Water Use By 90%. *Inhabitat - Sustainable Design Innovation, Eco Architecture, Green Building.*

Retrieved from http://inhabitat.com/poo-power-new-toilet-system-turns-human-waste-into-electricity-and-reduces-water-use-by-90/

Steinfeld, Carol and David Del Porto. (2007). *Reusing the Resource: Adventures in Ecological Wastewater Recycling*. Concord, MA: Ecowaters Book.

U.S. Geological Survey. (n.d.). Emerging Contaminants In the Environment. Retrieved from http://toxics.usgs.gov/regional/emc/

*Water*CAMP*us*. (2013). Retrieved from http://www.watercampws.uiuc.edu/index.php?menu_item_id=8

Water Sense. (2008). Indoor Water Use in the United States. EPA. Retrieved from http://www.epa.gov/WaterSense/pubs/indoor.html

WeBuildGreenCities. (2011). Retrieved from http://www.webuildgreencities.com/ [video]

Vestergaard-Frandsen. (n.d.). LifeStraw. Retrived from http://www.vestergaard-frandsen.com/component/docman/cat_view/51-product-brochures/57-lifestrawr

CHAPTER 5
RECYCLING AND ZERO WASTES
Part 1: Recycling

There is something satisfying about sorting the trash, pulling out the recyclables, and carrying them out to the curb on the appropriate day. It means we are doing our small part for the environment (Hawken, 1993). Recycling is essential in order to extend the life of materials, but NOT essential for those materials to just be made into a lower-level product. This is the lowest level of sustainability. This level says it is all right to consume as long as you discard appropriately (recycle). However, just because a material is recycled does not make it ecologically benign, especially if it was not designed properly in the first place, i.e., it was not designed to be recycled (McDonough and Braungart, 2002).

What was the world's largest manmade object? Was it the Great Wall of China or the Statue of Liberty? Actually it was a landfill in Fresh Kills, New York. One hundred feet (30 m) high and covering 4 square miles (10 km^2), it could be seen from space. Although it contained 2.9 billion cubic feet (82 million m^3) and 100 million tons (91 million t) of waste, it represented only 0.018% of the total waste generated in the United States at the time. We disposed of 5,500 times more solid wastes elsewhere in the United States (Hawken et. al., 1999). This particular landfill has now been closed and is being made into a park.

Many of our environmental issues are caused by our unending, ever-increasing desire to consume material goods and services. This desire to consume has been compared to a drug or alcohol addiction, and recycling to "an aspirin to alleviate a rather large collective hangover called overconsumption" (McDonough and Braungart, 2002). Our overarching goal should be not to recycle more, but to produce and consume less (McDonough and Braungart, 2002).

When plastics are mixed and melted together, they separate (such as when oil and water are mixed) and solidify in layers. These polymer blends have limited uses because the phase boundaries (layers) cause structural weakness (Polyethylene Recycling, 2012). Downcycling is a term given to products that can only be recycled into a lesser grade product, just delaying their journey to the landfill (Edwards, 2010).

What has come to be called "cradle-to-cradle" is the best action. First, because the item is made from a material that has a future life; it will at some point be made into a new product of at least an equal value, or has the potential to be made into a higher-grade product. Up-cycling prevents waste of resources by using existing materials over and over again (McDonough and Braungart, 2002) as the integrity of the materials is maintained (Edwards, 2010). Second, cradle-to-cradle means the product is designed from the beginning with no waste permitted. Any resulting byproduct is dealt with on the front-side, incorporating the so-called waste into another product line (McDonough and Braungart, 2002). There is no waste.

Our trash, or Municipal Solid Waste (MSW), includes table scraps, packaging, grass clippings and leaves, as well as larger items such as microwaves or sofas. From 1980 to 2005, MSW increased 60% (Williams, 2006-2013)! However, these numbers, reported by the Environmental Protection Agency (EPA) are beginning to decline, and from 2007 to 2009, MSW decreased from 255 million tons (231.3 million t) to 243 million tons (220.4 million t) (Williams, 2006-2013). Williams describes the EPA 2009 report with the following figures for those 243 million tons (220.4 million t) of solid wastes:

- 82 million tons (74.4 million t) were composted or recycled saving 33.8%;
- 29 million tons (26.3 million t) were incinerated to generate energy;
- 74% of office-type paper was recovered;
- 60% of yard debris was recovered;
- 34.5% of metals were recycled;
- Due to the recycling of 7 million tons (6.4 million t) of metals, greenhouse gas emissions were reduced (equal to removing 5 million cars from road for one year);
- Community curbside recycling programs are increasing;
- Only 3,000 community composting programs exist, a decrease from 3,227 programs in 2002;
- MSW sent to landfills per-person is lower than in 1960; and
- Due to population growth, the current total amount discarded MSW in landfills significantly higher than in 1960, yet lower than 1990 (Williams, 2006-2013).

So, although our waste per person is decreasing, we are still a waste-generating society. This will only change when we begin to change our habits and patterns.

These habits, patterns, and laws vary by state. In Oregon, the Beverage Container Act was put in place in 1972, and so consumers are quite used to the 5-cent deposit on most beverage containers (Campaign for Recycling, 2013). There are only 10 states that currently have a bottle bill: Oregon, California, Iowa, Michigan, New York, Massachusetts, Maine, Vermont, Connecticut, and Hawaii as well as Guam (BottleBill, 2007-2013). Detailed information regarding the facts regarding bottle bills is described online at Bottle Bill Myths and Facts (http://www.bottlebill.org/legislation/usa.htm).

Single-use plastic bags require the use of non-renewable fossil fuels (petroleum and natural gas) in their manufacture (Plastic Bags, 2009). They take thousands of years to biodegrade and they cause harm to wildlife. The many problems related to plastic bag usage are described by the Sierra Magazine (Plastic Bags, 2009). Many states have been trying to ban the use of plastic bags, but usage continues to be a struggle around the country. In Bellingham, Washington a model bag ban began August 1, 2013, whereas the Governor of Illinois vetoed a plastic bag recycling bill on August 27, 2013 (Campaign for Recycling, 2013). However, recycling at the local level is gaining momentum. According to National Conference of State Legislatures (NCSL):

- Currently, only 6 states—California, Massachusetts, New Jersey, Oregon, Rhode Island, and Washington are even considering banning single-use plastic bags;
- Hawaii has a plastic bag ban by county. If plastic bag usage does not decrease, fees in Hawaii could be 5 to 25 cents per bag;
- Hawaii, Louisiana, Maine, New Jersey, New York, Rhode Island, Vermont, and Washington may impose a tax on bags;

- Washington wants to establish rules for towns that choose to implement either a ban or fee on plastic bags;
- In 2013, Arkansas and Florida both failed to ban bags;
- Often, bag revenues in each state go to parks, school districts, community improvement trusts or other public programs; and
- Florida and Maryland proposed fee legislation, but the bills failed (NCSL, 2013).

According to Filters-Now (Top 5, 2012), the cities that were the best at recycling in 2012 were:

1. San Francisco, California: for having the best recycling programs, for exhibiting a very large scale of composting, and having a city goal of sending zero waste to landfills by 2020.
2. Boston, Massachusetts: the city will utilize a plant that will turn 50,000 tons (45.4 t) of leaves into reusable fertilizer or energy, 1.5 megawatt hours of power.
3. Chicago, Illinois: the Green Roof program is helping conserve water, filter rainwater, and improve energy efficiency.
4. Denver, Colorado: the Altogether Recycling Plant is focused on recycling glass, plastic, aluminum, tin, and steel, and paper.
5. Portland, Oregon: their focus is on being a part of the Total Reclaim Environmental Services, which recycles computers, electronics, fluorescent lamps, refrigerant gases and appliances.

Recycling varies from state to state, obviously. In Portland, Oregon, it is surprising to not see curbside recycling bins, whereas in Albuquerque, New Mexico it can be surprising to actually find someone who knows what recycling is, because no programs are available (Williams, 2006-2013).

Solid waste is an enormous problem. Many artists have decided to express their concern regarding this issue in the form of sculptures constructed from only waste materials. Stonefridge, or Fridgehenge, was a sculpture in Santa Fe, New Mexico, of more than 100 old refrigerators stacked up and arranged in a ring like the stones at Stonehenge in England. Strong winds have caused many parts of the 80-foot (24 m) high structure to tip over (Stonefridge, n.d.). The installment took much effort and time to construct in the late 1990's and it was dismantled in 2007 due to health and safety issues (Prattie Place, 2007) so the sculpture, created by Adam Jonas Horowitz, is no longer a cultural destination in the state. However, the construction

Figure 5.1. A replica of Stonehenge in England made up of discarded automobiles and tractors in Nebraska, called Carhenge. Source: Shutterstock.com.

efforts are documented at Prattie Place (2007). Another example of a Stonehenge type of display is in Alliance, Nebraska, called Carhenge, and is constructed from abandoned cars and tractors (Figure 5.1).

Developed countries produce 32 times more waste than developing countries (Edwards, 2010) (Figure 5.2). About a billion people live in the developed world (United States, Canada, Europe, Australia and Japan), whereas 6 billion live in the rest of the world (Edwards, 2010). Many have the goal to be like America. Let's hope they do not mean to waste as much as we do!

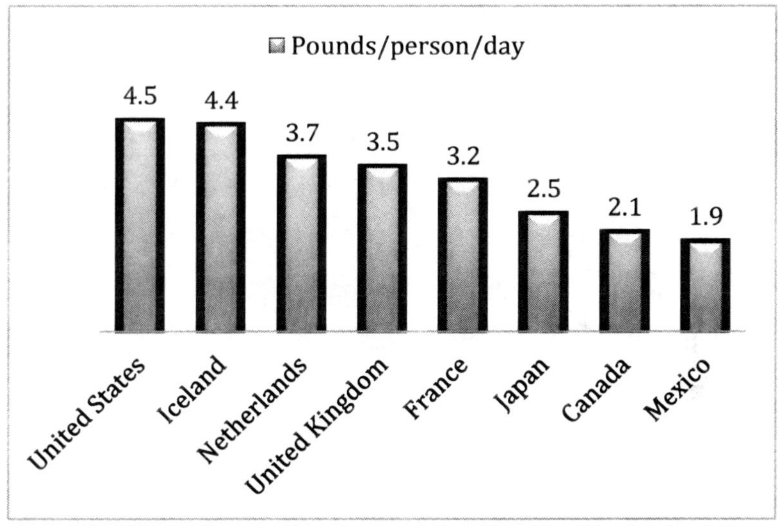

Figure 5.2. Waste Generation Around the World. The United States and Iceland alternate for first place in being the country that generates the most waste per person. Source: What the world eats and wastes (2013) Retrieved from: http://art4agriculturechat.wordpress.com/ 2013/05/02/what-the-world-eats-and-wastes/

Hawken et al. (1999) trace the manufacturing process of a typical aluminum can of cola-flavored drink in England (the process would be much the same for any country):

- Bauxite is mined in Australia. The bauxite is trucked to a reduction mill where it takes approximately 30 minutes to purify each ton (0.9 t) into a half ton (0.45 t) aluminum oxide;
- Giant ore carriers send the aluminum oxide to Sweden or Norway where hydroelectric dams provide cheap electricity. The ore sits at the smelter for two months;
- It takes 2 hours to turn that half ton (of aluminum oxide) into a quarter ton (0.23 t) of 10 ft long ingots of aluminum metal. Ingots are cured for 2 weeks and shipped to Sweden's or Germany's roller mills;
- The ingots are heated to 900 degrees F (°F) (482 degrees C [°C]) and rolled out to a thickness of 1/8 inch (3.2 mm);
- Sent to another country, rolled ten times thinner, they are now ready for fabrication;
- This aluminum is sent to England where it is formed into topless cans, washed, dried, and painted with a base coat and then product information. The cans are lacquered and sprayed inside with a protective coating;
- Cans are stacked on pallets and warehoused until needed;
- Cans are shipped to a bottler, washed and cleaned;
- Cans are filled with water, flavoring, phosphorous, caffeine, and carbon dioxide;
- Sugar is harvested from beets trucked, milled, refined in France;
- Phosphorous is excavated from open-pit mines in Idaho, which also unearths radioactive thorium;
- Cans are filled at a rate of 1,500 cans per minute and put in cardboard cartons;

- Cartons are made from forest pulp from Sweden or Siberia, or old-growth, virgin forests in British Columbia;
- Palletized cartons are shipped to regional distribution warehouses;
- Cartons of soda are bought by a supermarket within 3 days;
- Consumers buy the 12 ounces of phosphate-tinged, caffeine impregnated, caramel-flavored sugar water;
- Drinking the cola takes only a few minutes; and
- Discarding the can takes only a second (Hawken et al., 1999).

None of us fully grasp all that it takes to bring any of our stuff to us, or how easily we discard it. However, aluminum is one of those few "infinitely recyclable" materials with no loss in quality during the process. Of all of the aluminum produced in the United States, 31% is from recycled scrap, and 75% of all aluminum produced since 1888 is still in use today (Hawken et al., 1999).

In a cradle-to-cradle or restorative economy, every product (or by-product) is imagined even before it is made (Hawken, 1993). If the cost of mitigation is placed with the originator, that compels reasons to rethink the entire process and to avoid those challenging by-products (Hawken, 1993).

In Germany and Japan, the cost of taking a car to the landfill has a tipping fee (a charge to dispose of the car) that ranges from $300-$400 per ton ($273 to $364 per tonne), compared to American tipping fees of $20 to $30 per ton ($18 to $17 per tonne) (Hawken, 1993). An SUV can weigh 6,000 pounds (2,700 kg). In Japan and Germany, discarding that vehicle would cost about 900 dollars, but in the U.S., it would be only 60 dollars. Americans do not place enough value on our material resources.

Germany has also pioneered the concept of "extended product responsibility," which means "You make it, you own it forever" (Hawken et al., 1999). With that idea in mind, German designs incorporate easy disassembly and disposition. Their country has established post-user responsibility costs that are so high that companies design to avoid them. German recycling of packaging went from 12% in 1992 to 86% in 1997. During the same period, plastic recycling increased 1,790%! BMW now has a disassembly plant. Parts are bar-coded to identify materials and instructions for reuse. Their disassembly plant can strip a BMW Z-1 sports car in 20 minutes. Interestingly, the new part configurations result in easier repairs (Hawken, 1993). The systems BMW has designed are spreading across Europe and Japan (Hawken et al., 1999). We have to ask: just how long it will take to get to this side of the ocean, from either direction! Why is America so wasteful?

Electronic waste, commonly called e-waste, is growing at an alarming rate. E-waste contains toxic materials such as lead, mercury, cadmium, and brominated flame retardants (Total Reclaim, n.d.). Many of these chemicals bioaccumulate (become concentrated over time) and can cause severe birth defects (Total Reclaim, n.d.). How much e-waste is there out there? In the United States alone, each year we generate 3 million tons of computers, printers, phones, cameras, televisions, and refrigerators, and, globally, it is growing by 40 million tons per year (36.3 million t), "equivalent to filling 15,000 football fields six feet (1.8 m) deep with waste" (Vos, 2012), and it is increasing!

It is well known that China is the largest dumping ground for e-waste. By 2020, their e-waste from computers will jump by 200%-400% and mobile phones by 700% (Vos, 2012). Even worse are the numbers for India, where computer waste is expected to increase 500% and mobile phone e-waste will jump 1,800% (Vos, 2012). Recycling centers with state-of-the art facilities do exist, however, most e-waste is shipped both legally and illegally to developing countries where very primitive recovery systems are in place, with no health or safety regulations (Vos, 2012). Routine inspections of seaports in Europe in 2005 found about 47% of exported waste was illegal and that 25,400 tons (23,000 t) of e-waste was illegally shipped from the United Kingdom alone (Vos, 2012).

Some states have enacted TakeBack laws requiring manufacturers to take back their products for recycling and reuse (Figure 5.3). These programs have varied levels of actual success. Although Texas passed a TakeBack law in 2008 requiring manufacturers to have a recovery plan for computers, keyboards, and mouse devices at no charge to consumers, Texas ranks last in electronic recycling (Chen, 2010).

Of products purchased in North America, only 1% are still in use at the end of 6 months (Hawken et al., 1999). This opens up an enormous possibility to redesign all industrial production and uses for products, beyond recycling bottles and paper (Hawken et al., 1999). What opportunity presents itself to you every day? We need to rethink our relationship to everything we consume:

- What does it do?
- Where does it come from? and

- How can we get the service of this product from a net flow of very nearly nothing at all but better ideas? (Hawken et al., 1999).

It could be that businesses are missing a huge opportunity, as customers become more and more knowledgeable about the topic of sustainability. Instead of bombarding us with advertising to BUY, BUY, BUY, they should take advantage of the chance to inform us about their sustainability practices. If we were informed of what producers were doing to save the earth, we would naturally be more likely to purchase their product over another. But this approach requires an awakening in consumers and producers that is only just beginning.

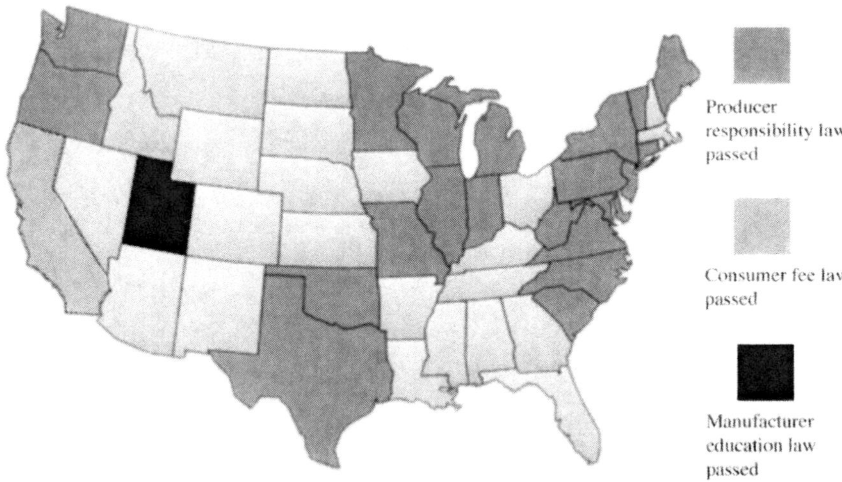

Figure 5.3. States that have legislation regarding e-waste disposition.
Source: http://www.treehugger.com/

Since we have learned that recycling is only the lowest level of sustainability, try looking at purchases from this perspective. Aim for zero: zero wastes, zero emissions, zero ecological impact. (McDonough, and Braungart, 2002). Instead of trying to be "less bad" what would it look like, what would it mean, to be 100% good? How might you do that? Aim for zero!

Part 2: Zero Wastes

To provide the average American family's needs for one year requires that industries move, mine, extract, shovel, burn, waste, pump, and dispose of 4 million pounds (1.8 million kg) of materials. Our average personal activities involve the mobilization of about 123 pounds (56 kg) of raw materials per day (Hawken et al., 1999)

Think about the waste around you. Your Chinese carry-out food may be kept warm in a Styrofoam carton for 15 minutes while you eat your lunch, but it could easily spend several hundred (or even thousands) of years in a trash heap attempting to decompose. To address this issue in Portland, Oregon, many food carts have adopted a system called the "Go Box." It is washed and reused and prevents excessive waste (Go Box, n.d.). It does not need to decompose. When the useful life of the box is complete, the container can be recycled.

What does it mean to decompose, to be biodegradable? We assume that the paper products and food scraps we put in our trash will at least biodegrade in a landfill. A University of Arizona study found, however, that that is not the case. In their research, they found "still-recognizable 25-year-old hot dogs, corncobs and grapes in landfills, as well as 50-year-old newspapers that were still readable" (Earth Talk 2014). In most landfills, there is very little oxygen, not a whole lot of soil, and very few microorganisms, all of which are needed to biodegrade the materials sent there. If these materials are exposed to air, the microorganisms will find their way to their food supply and begin the process of decomposition; but buried beneath many feet of other materials, encased in plastic and capped with clay as in most landfills, the story is different.

The term “lean thinking” was developed by Womack and Jones (2010). They define wastes as “any human activity which absorbs resources but creates no value.” This includes things such as mistakes, waiting unnecessarily, and unneeded processing steps (Hawken et al., 1999). Wastes used to be invisible, but now we are beginning to see them around us everywhere. For example, Hawken et al. (1999) describe the waste in the manufacture of glass automobile windshields:

- First, flat glass emerges from the furnace and is cut into pieces somewhat larger than the windshields;
- This is cooled, packed, crated, shipped 500 miles (805 km) to the fabricator;
- It is unpacked 47 days later and cut to shape (25% loss);
- It is reheated and formed to the correct curve;
- Next, it is cooled, repacked, shipped 430 miles (692 km) to a glass encapsulator;
- It is unpacked 41 days later, fitted with edge seals and other refinements, repacked, and shipped 560 miles (900 km) to the car factory;
- Finally, it is unpacked 12 days later, and installed in the car.

It has taken 100 days, and the glass has travelled 1,500 miles (2,400 km). None of this has given value to the customer (Hawken et al., 1999).

Every man, woman, and child in the United States produce twice their own weight in wastes everyday (Benyus, 1997). That is enough waste to fill two Louisiana Superdomes daily! That’s a lot of trash! At first, the earth was big and we were small. We didn’t notice when we began to foul our own nest as we moved into new territories, leaving exhausted soils and polluted waters in our wake. The data varies by source, but we add about 8 million people to the planet every month (World Population Balance, 2013), the same number of people that live in New York City (Benyus, 1997). Those 8 million people generate 3.3 million tons (3 million t) of trash each year (Lodge, 2010). In the U.S. alone, we generate 12 billion tons (10.9 billion t) of solid wastes every year; 200 million tons (181.4 million t) of wastes are added to the atmosphere each year; and 90,000 tons (81,647 t) of nuclear wastes are generated that will remain toxic for at least another 100,000 years (Benyus, 1997). We found *a* way to do what we needed to have done, but it was just *one* way, not necessarily the *best* way. We now have to find better ways to make what we want or need, and new ways that do not generate all of these wastes.

We reached a point several decades ago when we no longer had places for wastes. In 1987, a barge left Long Island and spent 6 months looking for a place to get rid of its cargo, 3,186 tons (2,890 t) of commercial garbage; no one wanted it. In 1988, another ship left Philadelphia and roamed the earth for 2 years looking for a place to dump its 15,000 tons (13,608 t) of toxic incinerator ash; the waste was dumped in an undisclosed location (Benyus, 1997). Where was the location, do you think?

Rachel Carson was the one who raised the first flag in the 1960s and 1970s with her book *Silent Spring* (Carson, 1962). Rachel Carson and her book started an environmental movement. In response, companies comply minimally with federal regulation, hiring lawyers to circumvent environmental protection laws. Industrial companies develop processes that generate wastes, and then spend 70 billion dollars each year treating and disposing of those wastes (Benyus, 1997). Penalties fail to deter companies that choose to pay fines rather than change their methods. What really does work, what gets designers in the United States back to the drawing board, is the greening of its customers (Benyus, 1997). If you care, businesses are forced to care as well.

Related to the natural world, the current niche for people is that of being opportunists, as we concentrate on growth and throughput (moving materials from the earth into products and transforming them into waste) (Benyus, 1997). We take advantage of abundance, and then move on. Other organisms that take advantage of abundance are weeds in an open field, bacteria in our leftovers, and mice in a predator-free barn (Benyus, 1997). Taking advantage of abundance is a Type I system. We entered this type of system at the beginning of the industrial revolution. Suddenly we had unlimited access to resources and we went crazy (Benyus, 1997). Just as with the law of thermodynamics, matter cannot be created or destroyed, and economic consumption is the same: it does not create or destroy matter, it only changes its location, form, and value. The same tonnage of mined resources is treated, transported, and made into goods that are distributed to customers, and then hauled away as waste or emitted as pollution (Hawken et al., 1999). We transform the abundance of the earth into wastes.

In the natural world, Type I species are the pioneers, the annuals that move into disturbed soils and set the stage for succession. They heal the earth (Benyus, 1997). The plants that are Type II species are the perennials; they make just a few seeds and then spend the rest of their energy building roots to get through the winter. The dominant species are the Type III species. They are loyal to the place where they live, staying in equilibrium with their environment; they do more with less, replacing anything that they take (Benyus, 1997). We have been acting like Type I species taking advantage of abundance. However, we prevent the environment from progressing through the stages of succession; we need to begin to learn from the Type III species who value their place, and use resources sparingly. Now that we have been everywhere on the planet, we need to close the loops to resource use, and not jump off to another planet to repeat our failed pattern (Benyus, 1997).

There are communities and companies that "get it." At an Ecopark in Kalundborg, Denmark, 4 companies moved within close proximity to each other because they are codependent on resources and energy (Benyus, 1997). Waste heat that cannot be used commercially is used to heat the homes of 20,000 residents (Indigo, 2003). In Europe, companies are required to take their packaging back, or hire another company to recycle it (Benyus, 1997). Packaging for shipping used to say, "This product can be recycled;" now it says, "We recycle our products and packaging!" (Benyus, 1997). In the future, product design must change so that products last a very long time, and can be disassembled easily for either recycling or reuse. Snaps will be used instead of glued joints, and parts made of only one material instead of twenty (Benyus, 1997).

We must make products that consider the reuse of materials right from the very beginning. Today's potato chip bag is 9 thin layers of 7 different materials (Benyus, 1997) making the bag impossible to recycle. A new bag needs to be designed that keeps the chips fresh, but the bag will be able to have a future beyond protecting chips (Benyus, 1997). Yes, the bags can currently be folded and woven into wastebaskets and purses, but do you have one? How many do you want?

Nature only makes what is needed exactly when it is needed. There is no waste. Where is nature's garbage pile? Where is nature's landfill? There is no such place. Nature does not waste anything.

Interface, by their own definition, is the world's largest designer and maker of carpet tile. But also for them, "design is a mindset and sustainability is the journey of a lifetime" (Interface, 2012). Every input of raw materials at Interface is considered to be waste until proven otherwise. Once they started to measure and follow inputs, they discovered that indeed most inputs were wastes (Hawken et al., 1999). Their goal is to be a company that generates no wastes of any kind, and in fact, has no impact on the earth at all.

Workers like to see waste eliminated, so whenever you get rid of wastes to create "lean" production, it makes people happier (Benyus, 1997). No matter where you are in the world, people feel best if their work "involves a clear objective, intense concentration, no distractions, immediate feedback on their progress, and a sense of challenge" (Benyus, 1997). Tasks that move companies in the right direction become an end in themselves. Traditional batch-and-queue production, which makes large quantities of products in anticipation of future orders, fails all of these criteria (Benyus, 1997). People need to have meaningful work to have value for their own lives. It is important to not waste people as well as resources.

Shipping pallets are used underneath all products transported around the world. Pallets use 11% of our total lumber, including 40% of hardwoods in the United States (Hawken et al., 1999). As there are 1.5 billion pallets in the United States, there are 6 pallets for every citizen. Hawken et al. reported in 1999 that broken pallets were rarely repaired, and the waste generated could frame 300,000 average homes. If just half of discarded pallets were reclaimed, 2,500 inner city jobs could be created, saving 765 million board feet (1.8 million m^3) of lumber, equal to 152,000 acres (61,512 ha) of timberland. Artisans do make use of 54,000 pallets per year to make furniture.

Trebilcock (2010) reported that 52% of companies buy their pallets locally, 41% use their pallets only 2 to 6 times and only 15% use them more than 20 times. Wooden pallets are greatly preferred, with 92% of companies purchasing them over any other material. When the pallet is damaged, 63% of companies use a pallet recycler because it is less expensive than repairing the pallet in-house. Pallet rental or recovery systems are used 39% of the time because the customers of the company require it. Of the primary changes that companies see in their pallet usage, only 3% to 5% expressed reasons related to sustainability or recycling. When asked what their company's potential is for participating in a pallet retrieval or recovery system, or in a third-party pallet rental system, 52% said they would or might consider it, and 44% either did not know what it was or would not consider it. It is apparent that change is happening over the past decade, but the fact that 44% were not on board shows just how slow change can be.

To prevent so much paper waste, Hawken et al. (1999) described the possibility of "decopier technology." Machines would remove toner from paper with so little marring that the paper could be reused 5 times. Immersed in water at 130° F (54.4°C) the ink would lift off the paper, be collected and reused. Although the ink would be expensive, it would never be thrown away. It could protect forests by reducing the need for paper by 90% (Hawken et al., 1999).

The delayed progress was summarized in an article by Friar et al. (2001). The new technology and potential tremendous demand caused complications in the research, finances, and operations (Friar et al., 2001). It has still been over a decade since this last report. I have not heard about any results from this project though. Have you? Why haven't we seen this product? Who would be harmed with its introduction?

Remember all the hidden history in the manufacture of the cola can? That hidden history is called an "ecological rucksack" (Hawken et al., 1999). For example, the waste generated to produce one semi-conductor chip is 100,000 times the chip's weight. A detailed description regarding the calculations behind an ecological rucksack is available online at http://www.gdrc.org/sustdev/concepts.html (Ecological Rucksack, n.d.). The calculations include the materials necessary for production, use, recycling and disposal of a product, but not the materials used in the product itself. Ecological rucksacks (ER) are calculated by subtracting the weight (W) of the product from the material intensity (MI) of the product or service: ER = W – MI. The MI has five categories: (1) abiotic raw materials; (2) biotic raw materials; (3) moved soil (agriculture and forestry); (4) water, and (5) air involved in the calculations (Ecological Rucksack, n.d.). The waste generated by production of a laptop computer is 4,000 times the laptop's weight. In order to bring 1 quart (0.9 L) of Florida orange juice to your table, 2 quarts (1.9 L) of gasoline and 1,000 quarts (946 L) of water are required; 1 ton (907 kg) of paper requires 98 tons (88.9 t) of other resources (Hawken et al., 1999).

Americans waste (or cause to be wasted) about a million pounds (454 t) of materials per person per year (Hawken et al., 1999), with the total annual wastes in the U.S. at more than 50 trillion pounds

(22.7 trillion kg) per year. Of this, less than 2% of the total wastes were recycled (Hawken et al., 1999). In a zero-waste society, this would be unacceptable. There would be no garbage because trashing anything would be unconscionable (Steffan, 2011). Striving for a world with zero wastes is closer than we think, and will make a world much better than we can imagine by improving the quality of the products we purchase, including the food we eat and where we choose to live. "A zero-waste life can mean a life bursting with better design and more quality" (Steffan, 2011).

We are all aware, to some degree, of the physical wastes around us. But there are other forms of waste that we ignore. We are the richest country in the world and yet we cannot:

- Balance the budget;
- Properly fund our educational system;
- Repair our bridges; and
- Take care of our infirm, aged, mentally ill, and homeless.

Where is our wealth going? Of the 9 trillion dollars spent each year in the U.S., at least 2 trillion is wasted. Waste has remained part of our outdated industrial system as the rest of the world advances. Waste includes highway congestion (100 billion dollars in lost productivity), accidents (150 billion dollars), and building and repairing roads (1 trillion dollars). We do not have the same efficiency standards as Japan and so lose 200 billion dollars a year in wasted energy. Polluted air costs us 100 billion dollars. Every year in healthcare, 65 billion dollars are spent on nonessential or fraudulent tests and procedures; 250 billion dollars are wasted on inflated and unnecessary medical overhead; 50 billion dollars are spent in health costs due to food choices; and 395 billion dollars goes to obesity, heart disease, strokes, and substance abuse (Hawken et al., 1999).

According to Hawken et al. (1999), we waste 250 billion dollars a year with our unnecessarily complex and unenforceable tax code in which we fail to pay the IRS 150 billion dollars per year. Crime costs us 450 billion dollars per year, and we spend 300 billion dollars on lawsuits. Hawken et al. (1999) note that the U.S. has 70% of the world's lawyers, and asks us to consider what the relationship might be to the number of lawsuits.

American nuclear weapons facilities have a clean-up bill of 500 billion dollars, not including the costs for the superfund sites or the 25 billion tons (22.7 billion t) of toxic, radioactive material wastes. The numbers ignore all the subsidies, government waste, consumer fraud, gambling, or replacement of shoddy products. Could it be possible that 50% of the Gross Domestic Product (GDP) is attributable to some form of waste (Hawken et al., 1999)?

Forty-five million tons (40.8 million t) of wastes are "wasted" each year (Hawken et al., 1999). However, farms are beginning to close these loops of wastefulness, and are reusing wastes to make ethanol. Many regions of the world could take better advantage of their unused resources, such as:

- A typical Nebraska harvest results in some below-grade grain or damp grain: enough to run one-sixth of the state's cars for a year (if they got 90 mpg [38.3 km per liter (KPL)] like the Hypercar);
- All the straw burned in France or Denmark could run all their cars year-round; and

- Ditto for nutshells in California, peach pits in Georgia, cotton-gin trash in Texas.

In the past, California burned its rice straw in the winter, resulting in lung disease downwind (Hawken et al., 1999). Now some of the farmers are flooding their fields after the harvest, turning the fields into a wildlife habitat for migrating birds. As the rice stalks decompose, the soil is rebuilt; ducks aerate and fertilize the fields; and hunters pay to visit the farms (Hawken et al., 1999). The crop yields increase and the extra income from the temporary wildlife habitat is a bonus.

With sales of over 400 billion dollars in 2008, Walmart is the largest corporation in the world (Edwards, 2010). Although it is still criticized for its business practices in wages, workers' rights and employee health, Walmart is making progress in other areas. It has 3 sustainability goals:

- Be supplied 100% by renewable energy;
- Create zero wastes by 2025; and
- Sell products that sustain our resources and environment.

Refer to an article by Clancy (2014) regarding the commitments Walmart has made with the members of its supply chain to meet their current sustainability goals.

A key step for a business to become more sustainable is for it to work with suppliers to encourage a green supply-chain. Walmart is developing a packaging scorecard to evaluate suppliers' sustainable practices. By working with just 1 toy manufacturer to reduce packaging on 16 items, Walmart used 230 fewer shipping containers. This saved 356 barrels of oil (56,600 L) and over 1,300 trees (Edwards, 2010).

In order to make a statement regarding our waste production, many are creating works of art to express concern. In 1996, artist H. A. Schult assembled 1,000 people made of trash, and has taken this enormous display around the world (Environmental Graffiti, n.d.). He ranks worldwide among the first environmental artists. Syaka Ganz turns trash into beautiful sculptures of animals (Ganz, 2014). Bernard Pras uses trash to reproduce great master paintings (Pras, 2014).

Sunday Afternoon on the Island of Grande Jatte, a beautiful painting by George Seurat (1884) hangs in the Art Institute in Chicago. The large 82-inch by 121-inch painting is in the style called pointillism, because the effect is created by millions of tiny dots of paint. Another artist, a photographer named Chris Jordan, created an amazing reproduction of Seurat's painting using the images of 106,000 aluminum cans, the number of cans that are discarded every 30 seconds in the United States (Chrisjordan, 2008). Works of art like these are powerful statements regarding waste in our country. Sometimes art can make a clearer statement when words are ineffective. Where do you see wastes in your city? How could you make a statement through art or music to develop greater awareness regarding this waste?

What does it mean to the individual to aim for zero wastes? For example, think about how you purchase meat. Do you select a pre-packaged amount of hamburger or chicken in a styrofoam tray sealed in plastic? Or do you walk another twenty steps and purchase your meat from the butcher who will wrap the selection in butcher paper? Butcher paper does frequently still have a plastic coating, but the environmental damage is minimal in comparison. It also helps to ask the butcher for a totally recyclable paper. That is the only way to get things to change.

The city of Baltimore posts a "10-step program to reduce your wastes" (Baltidome, n.d.):

1. Recycle: if you do not already do this, starting a "10-step program" may be a little challenging for you!
2. Bring your own shopping bag: take your reusable bags wherever you go, not just to the grocery store.
3. Avoid petroleum-based products. Only 50% of fossil fuels go into making gasoline. The rest is used to make products such as candles, lotion, shampoo, soap, makeup, detergent, toothpaste, perfume, and crayons (144 of 6000 items are listed on their site).
4. Donate unwanted goods instead of throwing them away.
5. Pick up someone else's waste (litter). The city of Baltimore, like most cities, has a huge budget for cleaning up wastes. The Bureau of Solid Waste spends more than $5,000,000 annually on litter clean-up.
6. Buy products with less packaging.
7. Stop junk mail.
8. Reduce your use of toxic chemicals.
9. Buy less plastic.
10. Consume less. According to CNN, the richest 20% of people consume 86% of everything sold for private consumption. The poorest 20% consume 1% of it. (Baltidome, n.d.)

What should be the goal for the individual when it comes to wastes? What should you add to an individual sustainability model regarding wastes? Should your goal gradually become "zero"? Don't let the task be overwhelming. It all starts with a series of "baby steps."

References for Chapter 5
Recycling and Zero Wastes

About.com. (2013). Do Biodegradable Items Really Break Down in Landfills? Retreived from http://environment.about.com/od/recycling/a/biodegradable.htm

Art 4 Agriculture. (2013). Waste Generation Around the World Retrieved from http://art4agriculturechat.wordpress.com/2013/05/02/what-the-world-eats-and-wastes/

Baltidome. (n.d.). 10 step program to reduce your waste, Retrieved from http://www.baltidome.com/Reduce_Your_Waste_Program.html

Benyus, Janine. (1997). *Biomimicry: Innovation Inspired by Nature*. New York, NY: HarperCollins Publishers Inc.

Bottle Bill Resource Guide. (2007-2013). Bottle Bills in the U.S.A., Retrieved from http://www.bottlebill.org/legislation/usa.htm

Bottle Bill Myths and Facts, from the Bottle Bill Resource Guide. (2007-2013). Retrieved from http://www.bottlebill.org/about/mythfact.htm

Campaign for Recycling. (2013). Bottle Bill Law in Oregon. Retrieved from http://www.campaignforrecycling.org/states/oregon/or_bottle_bill_law

Carhenge. (2014). Roadside America, Retrieved from http://www.roadsideamerica.com/story/2606

Carson, Rachel. (1962). *Silent Spring*. Anniversary edition (October 22, 2002). New York, NY: Houghton Mifflin Company.

Chen, Katherine. (2010). Texas Ranks Last in Recycling Old Electronics. Earth911. Retrieved from http://earth911.com/news/2010/05/28/texas-ranks-last-in-recycling-old-electronics/

Chrisjordan. (2008). What can be done with 106,000 aluminum cans? Series of three photographs. Retrieved from http://curiousphotos.blogspot.com/2008/05/what-can-be-done-with-106000-aluminum.html

Clancy, Heather. (2014). Walmart pushes for collaborative sustainability goals. GreenBiz.com. Retrieved from http://www.greenbiz.com/blog/2014/04/30/walmarts-push-collaborative-sustainability-goals

Earth Talk. (2014). Do Biodegradable Items Really Break Down in Landfills? Retrieved from http://environment.about.com/od/recycling/a/biodegradable.htm

Ecological Rucksack. (n.d.). Sustainable Developments: Concepts. Retrieved from http://www.gdrc.org/sustdev/concepts.html

Edwards, Andres. (2010). *Thriving Beyond Sustainability: Pathways to a Resilient Society*. Gabriola Island, BC, Canada: New Society Publishers.

Electronics Take-Back Coalition. (n.d.). State Legislation, Retrieved from http://www.electronicstakeback.com/promote-good-laws/state-legislation/

Environmental Graffiti. (n.d.). 1000 Trash People. Retrieved from http://www.environmentalgraffiti.com/news-trashpeople#M3ME40qicdfrXpgv.99.

Friar, J., R. Kinnunen and V. Aggarwal. (2001). Decopier Technologies, in *The Case Research Journal NACRA*—North American Case Research Association. Retrieved from http://www.thecasecentre.org/educators/products/view?id=77994

Ganz, Syaka. (2014). Reclaimed Creations. Retrieved from http://www.sayakaganz.com/

Go box. (n.d.). Doing Away with Throw Away. Retrieved from http://www.goboxpdx.com/ and http://www.katu.com/familymatters/go_green/GO-box-148405525.html?tab=video&c=y

Hawken, Paul. (1993). *The Ecology of Commerce: A Declaration of Sustainability*. New York, NY: Harper Collins Publishers Inc.

Hawken, Paul, Amory Lovins, and L. Hunter Lovins. (1999). *Natural Capitalism*. New York, NY: Little Brown and Company.

Indigo Development. (2003). The Industrial Symbiosis at Kalundborg, Denmark, Retrieved from http://www.indigodev.com/Kal.html

Interface. (2012). Retrieved from http://www.interface.com/

Lodge, Michelle. (2010). Trash Inc.: The secret life of garbage. Forget Finance, New York is Trash Capital of the World. CNBC. Retrieved from http://www.cnbc.com/id/39309851

McDonough, William and Michael Braungart. (2002). *Cradle to Cradle: Remaking the Way We Make Things*. New York, NY: North Point Press.

National Conference of State Legislatures. (2013). State Plastic and Paper Bag Legislation: Fees, Taxes and Bans; Recycling and Reuse. Retrieved from http://www.ncsl.org/issues-research/env-res/plastic-bag-legislation.aspx

Polyethylene Recycling. (2012). Plastics Recycling. Retrieved from http://www.perecoveries.co.za/index.php?pageid=2&level=recoveries_top1_content

Pras, Bernard. (2014). InspirationGreen. Retrieved from http://www.inspirationgreen.com/the-art-of-bernard-pras.html

Prattie Place. (2007). Stonefridge in Santa Fe. Retrieved from http://pratie.blogspot.com/2006/02/stonefridge-in-santa-fe.html

(The) Problem With Plastic Bags. (2009, May/June). Sierra Magazine. Retreived from http://florida.sierraclub.org/suncoast/documents/ TheProblemwithPlasticBags.pdf

Seurat, George. (1884). Painting: Sunday Afternoon on the Island of Grande Jatte. Retrieved from http://www.artic.edu/aic/collections/artwork/27992

Stonefridge. (n.d.). Farewell Stonefridge. Retrieved from http://www.nowpublic.com/farewell_stonefridge

Top 5 Most Recycle Friendly Cities in the U.S. (2012). Posted in Filters-Now. Retrieved from http://www.filters-now.com/news/top-5-most-recycle-friendly-cities-in-the-u-s/

Total Reclaim. (n.d.). The E-Waste Problem. Retrieved from http://www.totalreclaim.com/e-waste_problem.html

Trebilcock, Bob. (2010). Pallet Survey, What Moves Our Readers, Retrieved from http://www.palletcentral.com/files/MMH1009_SpecialRptPallet.pdf

Vos, Sophie. (2012, November 15). Electronic Waste Disposal. Nicholas School of the Environment, Duke University. Retrieved from http://sites.nicholas.duke.edu/loribennear/2012/11/15/electronic-waste-disposal/

Williams, Laura. (2006-2013). United States Recycling Statistics. Retrieved from http://greenliving.lovetoknow.com/United_States_Recycling_Statistics

Womack, James P. and Daniel T. Jones. (2010). Lean Thinking: Banish Waste and Create Wealth in Your Corporation. 2nd Edition. New York, NY: Free Press.

World Population Balance. (2013). Why is Population an Important Topic? Retrieved from http://www.worldpopulationbalance.org/faq

CHAPTER 6
WHAT WE EAT

To many of us, food is almost like a religion. Food shows where we are from and who our ancestors were. It describes how we were raised, and how and why we gather around the table (NWEI, 2009). It is important to know where our food comes from, so that we are certain that it was produced in a clean way; that its production does not harm the environment; and that the workers received fair compensation for their work (Edwards, 2010). What does it mean when a label says "natural," "Grass-fed," or "Cage-free?" The USDA (2012) defines the terminology we find on our food labels. This is necessary in order to prevent "green washing," which is what happens when companies make inaccurate sustainability claims (Edwards, 2010).

There are about 200,000 species of plants. However, 75% of our food crops come from only 7 plant species: wheat, rice, corn, potatoes, barley, cassava, and sorghum. If you add soybeans, sweet potato, sugar cane, sugar beet, and banana to the list then that accounts for 80% of total crop tonnage. This use of only 7 crop species transforms complex ecosystems with thousands of species into ecosystems with only one, a monoculture (Hawken et al., 1999).

If you go to a standard grocery store, fresh fruits, vegetables, meat, and dairy are along the outer aisles of the store. This is the area of the store where most of our purchases should take place. Most of the food in the middle aisles is processed food, and the negative impact this food has on our health, as well as the health of the planet, has increased over the past 50 years (Center, 2008). We transport more than 817 million tons (741 million t) of food around the planet each year, more than 4 times what it was in 1961 (Worldwatch, n.d.). Processed food travels an average of 1,300 miles (2,092 km) before it gets to our table (Center, 2008). In an Iowa State University study, food that came from local farms travelled an average of 44.6 miles (71.8 km), whereas, if from conventional national producers, the same groceries would have travelled 1,546 miles (2,488 km) (National Geographic, 2008). Cheerios? They travel 1729 miles (2,783 km) to get to Portland, Oregon, from Minneapolis, Minnesota. Johnsonville Sausage? Sheboygan Falls, Wisconsin, is the location of their processing plant, 2,034 miles (3,273 km) from Portland. Surely there must be an option that does not require so many "food miles" (the distance food must travel to reach us). If not, then we have to begin to realize that this is a business opportunity! Each state needs to begin to produce its own products and have a local brand. Portland has a local brand of mustard, McMenamins Spicy Dijon Mustard and Terminator Stout Mustard (McMenamins, n.d.), and its own organic ketchup (Portlandia, 2010). That is a start, Portland. Keep it going! What about *your* city?

Transporting food requires fossil fuels to get the products to you. One problem with the high mileage numbers involved in food transportation is that it assumes that the current transportation system will always be available to us. It causes us to grow beyond what the land can support, a factor that contributed to the downfall of several empires (Benyus, 1997). We have lost our connection to where our food comes from. Our appetites need to be reconnected to our local environment (Benyus, 1997). In Nepal, they have 6 seasons and 6 diets, 1 diet for each season. If a food is not in season, it is not available. We should reconnect to this type of diet as a part of our rebalancing.

Much of our food in the United States also has (unfortunately) Genetically Modified Organisms (GMOs) inserted into its DNA. This is done to create plants more resistant to pests, cold, and drought, as well as to increase yields. Since the 1990s, GMOs are now present in more than 60% of processed foods (all those foods in the middle aisles), and no labeling is required in this country. Health risks were not known until recently, but allergies and increased risk of cancer are now understood to be caused in part by GMO foods (Adams, 2012). There are hundreds of articles about GMOs. Most are not complimentary. Many countries have restricted, labeled, or banned GMO crops. Sixty-one countries (40% of the world's population) already label genetically engineered foods (LabelGMOs, n.d.; Gucciardi, 2012) (Figure 6.1) including every country in the European Union. Genetically engineered foods are even labeled in China (LabelGMOs, n.d.).

GMOs can also create "superweeds" as the specialized genes can jump from the crop plant to weeds (Brown, 2005). Contrary to claims that genetically-engineered crops are reducing chemical applications, GMO plants have caused a substantial increase in the number and amount of herbicides needed (Benbrook, 2012), resulting in farmers applying even more chemicals than before. Glyphosate and herbicides such as Roundup® are actively toxic to human cells

(Baudouin, 2013). In a study at the University of Caen, France, glyphosate was found to cause moderate to severe liver damage in fish and aquatic life. The application of excess herbicides to farms runs off into groundwater and seawater. Glyphosates have even been found in urine of city residents (Baudouin, 2013). This shows that the chemicals persist through the food chain.

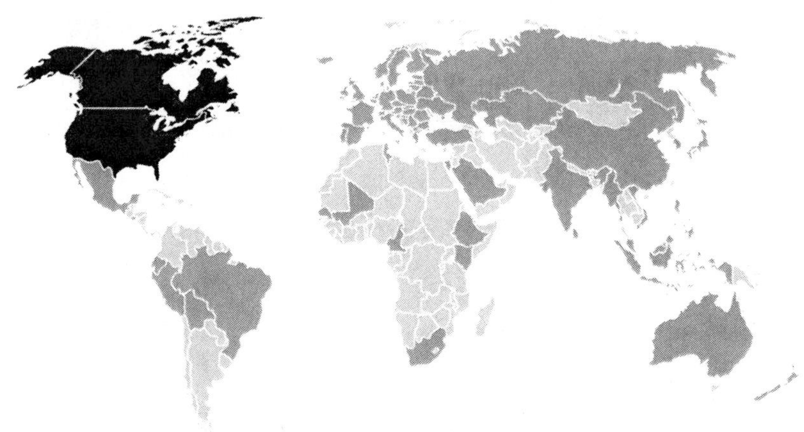

Figure 6.1. Sixty-four countries that have laws regarding GMO crops (medium gray) and the only two major countries with no laws (black). Source: http://justlabelit.org/right-to-know/labeling-around-the-world/

Becoming more aware of what we eat and where it comes from is a necessary trend if we are to become sustainable and in balance with our environment. We are being asked to think about our food. Do you get certified eggs or cage-free? Are vegetarian-fed chickens better? What are Omega-3 eggs? Is there a difference between brown and white eggs? Is local best, or organic? Cheapest? How do you decide? If you cannot decide, do you just call your housemate or spouse and ask, “Can you pick up some eggs on the way home from work?” That relieves you from the job of deciding (NWEI, 2009), but we have to remember that there are no perfect answers; we have to make tradeoffs because the systems are complex. After all, eggs are perplexing. But if you find yourself asking questions, you are on the right path (NWEI, 2009).

Where do you get your food? The local grocery store? Farmer's Market? Do you grow your own? Do some research to find the best farmer's market or store. You may be surprised to find just how much your favorite shop is doing to become more sustainable. New Seasons Market is a local chain in Portland, Oregon, with a comprehensive sustainability plan. After a decade, they are still local and have 3 branches to their plan: 1) Take care of the environment; 2) Take care of the foodshed; and 3) Take care of their staff. Whole Foods Market, a national chain, was challenged to do the same in "The Omnivores Dilemma" by Michael Pollan. This began a conversation between that author and Whole Foods' CEO to bring positive changes to both Whole Foods' operations and the food industry as a whole (Edwards, 2010).

Some of us are gradually making the transition back to organically grown produce. But even organic is not enough now. We need to know exactly how was it grown. Where was it grown? Is it in season here? Who grew it? At what scale? The transition to sustainable food means we grow our own food or purchase food that was grown no more than 100 miles (161 km) from home. The term "locavore" was coined by the residents of San Francisco to promote the purchase of food grown within a 100-mile radius. We expect whatever we want to eat to be available at the local grocery store, year round. Our physiology is actually not adapted to a diet of unlimited choices, but it will take us awhile to reconnect back to eating local food. As a starting point, think about getting everything from within a 200-mile (322-km) radius. For us in Portland, Oregon, that means getting our food from the Pacific Northwest (Washington and Oregon). Our fruit can come from The Dalles and Washington, our seafood from the coast, and wine, vegetables and dairy from the Willamette Valley. We are in a great location. What can *you* obtain from within a 200-mile radius?

A Community Supported Agriculture (CSA) provides individuals with produce on a weekly basis grown at a local farm. This simplifies getting fresh, seasonal produce for individuals, and guarantees an income at the beginning of the season for farmers. Between 1990 and 2008, CSAs have increased in number from about 50 to 2,500 in the Northwest, and to 13,000 across the country (McFadden, 2012).

While CSAs support individuals and communities, businesses can also have their needs met by the local environment. BALLE, or the Business Alliance for Local Living Economies is an international network of 15,000 businesses who want to buy local or fair-traded food. BALLE believes "*real* national prosperity — even global prosperity — begins at the local level, and that by connecting entrepreneurs who are re-thinking their industries, funders who are investing in the local economy movement, and network organizers who can mobilize on a broad scale, we can — and will — create a stronger, more resilient, and fair economy" (BALLE, 2012). When we can meet our needs locally, we have greater control over our lives (Edwards, 2010).

Why is it important to eat organic food? If we eat organic, we avoid consuming a host of chemical residues from pesticides, herbicides, and fungicides. Organically grown produce protects the environment, waterways, and other organisms since it avoids the introduction of unnatural chemicals. Organically grown food has higher nutrient values (Organic Trade Association, 2011). Diversity is preserved, and soil can rebuild instead of erode. Rural communities remain healthy when urban areas purchase locally grown produce (National Geographic, 2008).

What does "organic" on a label mean? In order to have the "100% Organic" label, all the ingredients must be 100% organic. If the label says "Organic," only 95% of the ingredients have to be organic. If the label says "Made with organic ingredients" it means that at least 70% of the ingredients are organic (National Geographic, 2008; FDA, n.d.).

We think that in switching to organic foods, that the same items will cost more; that is not always the case. There are five easy ways to go organic. Choose just a few foods, the ones that you eat most often, and switch those foods to organically grown versions. That will have a great impact on your diet and chemical load. For example, pesticides have been found in 30% of conventional milk (Parker-Pope, 2007). About 30% of our vegetable consumption is in the form of potatoes, one of the most pesticide-contaminated vegetables. Even after washing and peeling, 81% of potatoes still contain pesticides. Conventional farmers grow most of the peanuts used in our peanut butter; 99% of those farmers use fungicides to treat the peanuts for mold (Parker-Pope (2007). Organic ketchup has double the antioxidants of conventional ketchup (Parker-Pope, 2007) as well as fewer pesticides. Apples are the fruit we eat most often after the banana, and apple juice is our second favorite juice, with orange juice being the most popular. Most people do not know, however, that apples are the most pesticide-contaminated fruit (Parker-Pope, 2007). Switch to eating organically grown versions of these five foods, and you have substantially reduced chemical exposure to your body.

The "Dirty Dozen" and the "Clean Fifteen" are also a good place to start (Figure 6.2). The dirty dozen are fruits and vegetables that, even after thorough washing, still contain pesticide residues. The Environmental Working Group (2013) tests all fruits and vegetables each year to determine which ones are "clean," or safe to

The Dirty Dozen (Always Buy Organically Grown)

1. Apples
2. Celery
3. Cherry Tomatoes
4. Cucumbers
5. Grapes
6. Hot Peppers
7. Nectarines (imported)
8. Peaches
9. Potaoes
10. Spinach
11. Strawberries
12. Sweet Bell Peppers
    +PLUS
    Collards and Kale
    Summer Squash and
    Zucchini

The Clean Fifteen (OK to Buy Conventionally Grown)

1. Asparagus
2. Avocado
3. Cabbage
4. Cantaloupe
5. Corn
6. Eggplant
7. Grapefruit
8. Kiwi
9. Mangos
10. Mushrooms
11. Onions
12. Papayas
13. Pineapples
14. Sweet Peas (frozen)
15. Sweet Potatoes

Figure 6.2. Dirty Dozen and the Clean Fifteen (2013). Source: Environmental Working Group. Retrieved from http://www.ewg.org/foodnews/list.php

eat if grown conventionally, or "dirty," and therefore recommended to eat only as organic. In some cases, 15 different pesticide residues have been found on a single sample (EWG, 2013). The clean fifteen are fruits and vegetables least likely to hold pesticide residues (EWG, 2013). The lists change slightly from year to year, but if your organic purchases always include the dirty dozen foods, plus the five easy choices listed above (milk, potatoes, peanut butter, ketchup, and apples), and whatever else you consume in large amounts, you will make a dramatic difference in reducing the chemicals you consume and provide to your families.

If you have to choose, which is better, organic or local? Organic foods are cleaner, leaving no chemical residue in the environment, thus protecting all species. Organic is a good choice even if the food-miles are high (NWEI, 2009). Local is good, even if it is not organic, because the food miles are low and local economies are supported. Best of all, of course, is both organic AND local (NWEI, 2009). Studies continue to debate the issue, and it still must be an individual choice based on current knowledge (Watson, 2012).

Some individuals' immune systems are so sensitive (so sensitized from an accumulation of chemicals), that their bodies can not bear the burden of any more chemicals. Those people must eat organic food all the time; it is a matter of life and death to them. For those people, organic is essential, and the number of those individuals is growing. Clean food means the environment gets cleaned at the same time. We have very little understanding of the impact chemicals have on our systems, and the sensitivities that can develop as a result. Reed (2005) describes the conditions when a person develops strong negative reactions to everyday chemical exposures, a condition called multiple chemical sensitivity (MCS). This includes chemicals in our food, air, water, and even electromagnetic shifts in our surroundings (Reed, 2005). It is important to reduce the chemical load to which we are exposed; it can start by growing and buying organically grown produce.

Actually, organic is not the final word: biointensive mini-farming may be next. Biointensive farming combines deep cultivation to improve root growth with composting crops, optimizing microclimates by placing plants close to each other in wide beds, and mixed species planted together to deter pests naturally, because pests can't find their target plant (Hawken et. al., 1999). Minimal work is required because nature takes over. Startling results show high yields and much higher nutrient content for food farmed in this way (Hawken et al., 1999).

If you are not convinced and you still need another reason to go organic, look to the fact that in industrial nations such as our own, our bodies are, technically speaking, too toxic to be placed in a landfill (Hawken, 1993). Our bodies bioaccumulate toxic chemicals and contain high enough levels to be unsafe for disposal in a public landfill. Luckily, we have a place where we have decided this is acceptable, and they are called cemeteries! The sustainable options for "body disposal" at the time of death is another topic!

The Northwest Earth Institute (2009) says that to eat responsibly requires that we find ways to grow our own food; that we prepare and eat it ourselves; and that we compost the scraps. Use the rule of the "Locavore" and buy the food that is produced closest to your home. Get to know the farmer or orchardist who raises your food for you (NWEI, 2009). These are all great options to get started on living sustainably.

Also, don't ignore the life histories of the animals you choose to eat. Take some time to understand the industrial food production system (NWEI, 2009). Consider the stress chemicals released by the body of any animal raised under factory farm conditions: these conditions are generally over-crowded, unsanitary, and traumatic.

These commercial environments are commonly called a Confined Animal Feeding Operation or CAFO (See Gurian-Sherman [2008] for details). Given the alternative of a peaceful, clean, stress-free environment (pasture-raised and fed), which environment would you trust to produce the meat and eggs for you and your family?

Another reason to consider growing your own food and to develop local markets has to do with emergency situations. Most homes have a 3-day supply of food on hand at any given time. Most local stores have 3 days worth of food to supply their customers. Most distribution centers have 3 days of food to supply the stores they service. In other words, there is only enough food in your local community right now to last 9 days. In any emergency, the grocery store shelves empty within hours.

According to Rotenberg, there are thirteen things you can do to support a sustainable food system.

1. Eat foods grown and processed locally. Grow some of your own food.
2. Give preference to organically grown.
3. Buy from farmers' markets, food co-ops, and CSAs.
4. At supermarkets, ask where the food comes from. For example:
    a. Stores that do label: Whole Foods, New Seasons, Fred Meyer;
    b. Stores that currently do not label country of origin: Safeway.
5. Eat foods that have been processed as little as possible.
6. Buy food in bulk to avoid excessive packaging.
7. Compost food scraps and wastes.
8. Eat lower on the food chain (eat less meat).

9. Look for "fair-trade" on labels, if buying imported foods.
10. Cook at home as much as possible, or patronize sustainable restaurants.
11. Support practices that reduce negative environmental and social impacts by paying a little more for organic, if necessary.
12. Teach family, friends, and community about food choices and their impact on the environment
13. Pat yourself on the back for doing a good job, for doing what you can to do your best. Take pride in the fact that you're making the best decisions you can. Every transition takes time (Rotenberg, 2009).

Be creative and find a way to have a garden. A house I rented recently had mostly concrete for the back yard. So I purchased kids' plastic swimming pools, bought the best soil I could buy, and planted all kinds of vegetables (Figure 6.3). It was astounding to see how much I could harvest from 6 little swimming pools! I have seen images of these swimming pool gardens in cities across the country from Portland to Chicago, all providing an abundance of fresh vegetables. Rooftops, vertical walls, and reclaimed junk yards are being converted to organic farms (Walljasper, 2007). Raised beds full of vegetables are showing up in the strip of land between the curb and the sidewalk. Balconies are overflowing with containers of vegetables and flowers. There are options for a small garden everywhere. We just have to start looking.

Meat consumption is another concern. How much do we need versus how much do we eat? In the United States, we eat about 8 ounces (227 g) of meat per day, whereas the amount that we need is only 2 to 3 ounces (57 to 85 g) per day. The Mayo Clinic (2013) states that even 4 ounces (113 g) of red meat daily increases the likelihood of death 30% from *any* cause; this means that whatever health issues you have, if you eat 4 ounces (113 g) of red meat daily, you are 30% more likely to die from that health issue than someone who eats chicken or fish. It also takes 2.2 to 20 pounds (1 kg to 9.1 kg) of feed grain to make one pound (454 g) of meat. The amount of meat we traditionally eat in this country is out of balance with what the rest of the world eats (Figure 6.4).

Figure 6.3. "Swimming pool gardens" with cucumbers, squash and tomatoes. Source: Linda Pope.

If just 1,000 Americans ate 1 beef meal less per week, 70,000 pounds (31,751 kg) of grain could be saved; we could prevent 70,000 pounds (31,751 kg) of topsoil from eroding; and 40 million gallons (151.4 million L) of water could be saved (National Geographic, 2008). It takes 2,464 gallons (9,327 L) of water to produce one pound (0.45 kg) of beef in California (data from the Water Education Foundation). The same amount of water could be used to take a 7-minute shower every day for 6 full months. It only takes 25 gallons

(95 L) of water to produce a pound (0.45 kg) of wheat (Ogden, n.d.). All of those savings result just from the decision to eat one vegetarian meal per week. If you replace red meat and dairy with chicken, fish or eggs one day per week, greenhouse gases would also be reduced (equivalent to driving 760 fewer miles (1,223 km) per year). Switching to vegetables for that 1 day per week reduces greenhouse emissions equivalent to 1,160 miles (1,867 km) of driving (Center for Environmental Education, 2008). If you are not quite ready to give up meat, then the best option is to buy 100% organic or grass-fed animals, and to buy from local small farmers (National Geographic, 2008). Don't assume the terms "Free range," "No antibiotics administered," "No hormones administered," or, especially, "Natural" mean that a product is better. The FDA has approved these terms, but there are no third parties that verify the information (FDA, n.d.; National Geographic, 2008).

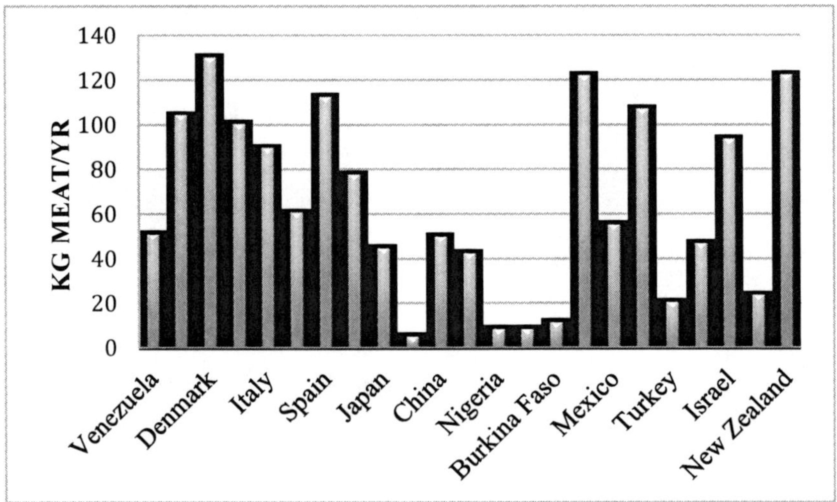

Figure 6.4. Annual per capita consumption of meat. Data shown for 2000. Many countries have similar patterns, by continent. Source: Food and Agriculture Organization of the United Nations (FAO), FAOSTAT on-line statistical service (FAO, Rome, 2004).

Another way to reduce your impact is to raise some of your own protein. Consider raising chickens for eggs and, potentially, for meat as well. The City Chicken (n.d.) has much of the information you might need to get started. Many people in Portland also raise goats for milk and cheese options. Rabbits are an easy source of protein, and fish (aquaculture) are also popular. If you raise your own source of animal protein, and you have known the animal all of its life, when you do have meat, it is with gratitude and greater understanding of the sacrifice made by the animal for you (NWEI, 2009).

How we raise animals at a large scale has to change. A study showed that very large herds of native African ungulates (hoofed animals) do not harm the ecosystem (Hawken et al., 1999; The Savory Institute, 2014). When local tribes herd cattle, only moderate damage is done to the ecosystem. But when foreign ranchers do the herding, the ecosystem is destroyed (Hawken et al., 1999; The Savory Institute, 2014). The researchers discovered the need to have animals graze more intensely for shorter and less frequent periods of time to improve the carrying capacity and health of the land (Hawken et al., 1999; The Savory Institute, 2014). As a result, ranchers are changing to a Holistic Management system (The Savory Institute, 2014), or to the Management Intensive Rotational Grazing (MIRG) method, representing the fastest-growing farming practices. Slightly less milk is produced by the free ranging cows over the confined ones, but it is at a lower operating cost, so income is actually higher per cow. Depositing their manure as they roam from area to area, grasses can recover to improve the habitat and wildlife. The health of the cattle and purity of the milk also improve, and there is no run-off from the open fields, so water is not polluted (Hawken, et. al., 1999; The Savory Institute, 2014).

It also helps to understand the whole backstory of what we eat as well (Refer to chapter 12 regarding backstories). It turns out that local is not always best. In New Zealand, they are committed to energy efficiency on their grazing lands, and have some of the most productive grazing lands in the world (Steffen, 2011). The British eat a lot of lamb, and when it's shipped from New Zealand, halfway around the world, the emissions generated by the lamb raised in New Zealand was 25% of that for the lamb raised in England. So it is not how far the product travels in every case, but the total energy used from production to your plate, of which transportation is but one part (Steffen, 2011). This is what understanding the backstory of your products means.

One aspect of living sustainably is to care about every person on earth. Many farmers in the Global South (Africa, Central and Latin America, and most of Asia) depend on being able to export their crops to generate the income needed for them to survive. By totally avoiding the purchase of crops grown non-locally we can be contributing to those farmers' continued impoverishment. A more ecologically sound route would be to make sure the item exported is fair-traded (Steffen, 2011). Products that bear the Fair Trade logo guarantee that farmers and workers are justly compensated for their work. Fair Trade USA helps farmers to build sustainable businesses in developing countries; this action provides positive influences in their communities (Fair Trade, 2010), so Fair Trade products help uplift entire communities. There is a balance to consider between supporting the communities of the Global South and supporting local farmers.

In 2008, it was predicted that all fish species would reach collapse (have less than 10% of their original number) by 2048 (National Geographic, 2008). A report published in 2013 tells us that many species are returning to abundance since the 1996 catch restrictions went into place (Wines, 2013). If we are to continue this trend, it is essential that we purchase only sustainably caught fish. Look for the Marine Stewardship Council certification or the "Fish Forever" seal. Other labels that can be trusted are listed below (Figure 6.5).

Although some fish are recovering, there are others that are slipping further into decline, the bluefin tuna being a good example. Walsh (2013) says the statistics were very difficult to find, but the International Scientific Committee for Tuna and Tuna-like Species in

Bird Friendly	Free-Farmed
American Human Certified	Grass-Fed
Demeter Biodynamic	Marine Stewardship Council
Fair Trade Certified	Rainforest alliance
Food Alliance Certified	Salmon-Safe

Figure 6.5. Labels that can be trusted (certified by a third party).
Source: The National Geographic (2009).

the North Pacific Ocean released their assessment in January, 2013, and stated that we have removed 96.4% of these animals from the ocean leaving only 0.036% of their original numbers remaining. The same week that this assessment was published, "a 489-lb [222 kg] bluefin tuna was sold at a fish auction in Tokyo for a record $1.76 million—or about $3,600 per pound" ($7,929 per kg) (Walsh, 2013).

The question arises: how do you reach people and inspire them to become sustainable when those with enough money will do anything for their prized bluefin sushi? Is there not a way to criminalize these activities worldwide? It is not just the Japanese that cause the problem. In 2008, the European Union fought to keep the bluefin quota at 22,500 tons even as the International Commission for Conservation of Atlantic Tuna recommended that the limit be set at 7,500 tons (6,804 t) to protect existing stocks (McDermott, 2009). For 2014 and 2015, the limits have been set at 14,881 tons (13,500 t). Each country has a specific limit. For example, Canada will have 532 tons (483 t) as its share of the 14,881 tons (13,500 t) (ICCAT, 2014), or about 1,932 bluefin tuna each year.

The projections by McDermott of "3 years left" in 2009 were essentially correct now that 4 years have passed and the numbers are below 0.4% remaining. McDermott (2009) correctly assessed the situation when he wrote, "Don't sell it. Don't buy it. Don't eat it." If you are going to eat tuna, be sure to select "chunk light," which is usually yellowfin or skipjack tuna, and which reproduce faster than other tuna species (National Geographic, 2008).

Although there are 15 species of tuna, MacKenzie (n.d.) writes extensively about the 5 key tuna species. There are many issues regarding how tuna are caught, as many other endangered species are caught in the nets simultaneously (called bycatch). Skipjack makes up 70% of our American tuna market. Whenever possible, look for pole-and-line caught skipjack (MacKenzie, n.d.). For yellowfin, eat it sparingly and again look for pole-and-line caught to avoid indiscriminate bycatch. Most populations of albacore have been overfished, or are in decline. Eat only pole-and-line caught Pacific albacore. Avoid bigeye, a tropical tuna, as it is in serious decline. Bluefin should be totally avoided. If caught, it should be sold only in very expensive sushi restaurants. In this case, it may be called "o-toro" on the menu (MacKenzie, n.d.).

When considering salmon, which is better – organic, wild-caught, or farmed? Actually, the thing that matters most is whether the fish is fresh or is frozen. Frozen salmon has half the environmental impact that fresh salmon does, due to storage during transport (Steffen, 2011). If 75% of Japanese consumers were to switch to frozen salmon, the ecological benefit would be greater than if all Europeans and North Americans ate only locally farmed or caught salmon (Steffen, 2011).

In many countries, fish are imported to meet the protein needs of its residents. Ghana is one of these countries. However, aquaculture could meet Ghana's entire need for fish (Aquaculture, 2013). There are many systems described online, including do-it-yourself systems for business (Burdette, n.d.) and individuals (Schoepke, n.d.). It is possible to set up a tank in a basement and grow enough fish to cover the expenses of your rent or mortgage. Tanks can accommodate several hundred fish and systems are designed to produce nearly zero wastes (Steffen. 2011). Tanks can be linked to crops to incorporate fish waste products as fertilizer for the plants (Burdette, n.d.).

Chocolate is sometimes associated with being a "forbidden food;" the Mayans considered it "a food for the Gods" (Yale, n.d.). The cacao plant grows naturally in African, Indonesian, and Brazilian rainforests, growing in the shade of these forests (National Geographic, 2009). Unfortunately, because cacao yields are highest in full sun, natural forest habitat has been destroyed, and the requirement for pesticides and fertilizers has increased. Residues of pesticides and fertilizers are found in chocolate (National Geographic, 2009). To protect rainforests and yourself, buy "Certified Organic" or "Rainforest Alliance Certified.

Although The Hershey Company is making progress and plans to source 100% certified cocoa for many of its products by 2020, the products sold at cash registers in every grocery store across the nation are not among the products listed. Hershey is well known for its poor labor practice of purchasing cocoa from countries that use forced-child labor in their fields, including 1.8 million children in the Ivory Coast and Ghana. Involuntary child labor is common in West Africa, especially in the Ivory Coast where Hershey sources its chocolate (Robbins, 2010). Children, ages 11 to 16, are forced to work up to 100 hours a week with no pay, very little food, and often abusive conditions. Hershey manufactures 80 million Hershey's Kisses a day using cocoa that may have come through the abuse of children (Hals, 2012; Robbins, 2010). Guarantee children's well-being by looking for the "Fair Trade Certified" logo (National Geographic, 2008). Robbins (2010) lists the companies that have switched to Fair Trade Certified chocolate.

A lawsuit has been filed against The Hershey Company, claiming that the company "violated anti-trafficking laws and knowingly benefited from a supplier using child labor" (Hals, 2012). A month after the suit was filed, Whole Foods Markets decided to drop the Scharffen Berger brand of chocolate (a subsidiary of The Hershey Company) after pressure from activists sought to highlight the issue of child labor in chocolate production (Hals, 2012). It does not take much. A handful of letters from questioning customers promoted this corporate decision. One of my previous student's eight-year-old son stopped eating Hershey's chocolates when he heard about the practice, and even years later, he has not purchased any of their products. A letter is more powerful than you can imagine. It should be everyone's number-one method to start change from within your home.

Coffee is another product we must purchase from tropical countries. Coffee plants thrive in the shade but produce 5 times more beans in the sun (National Geographic, 2008). With the removal of the trees, there are 97% fewer bird species now present. Previously, leaves would naturally drop from the trees, providing nutrients to the coffee plants; now fertilizers must be used (Craves, 2006). Weeds thrive in the sun, so now herbicides are used to control them. Tropical rains erode the topsoil increasing the need for even more fertilizers (Craves, 2006). So, just to survive, the sun-grown plants need pesticides, herbicides, and fertilizers. The use of these chemicals exposes workers to toxic chemicals on a daily basis (National Geographic, 2008). The best thing that you can do is to look for "Certified Organic" and "Fair Trade Certified" labels, or "Shade Grown," "Bird Friendly" or "Rainforest Alliance" on your coffee labels to guarantee that these ecosystems are protected.

Chemical solvents such as methylene chloride (a possible carcinogen) and ethyl acetate (which may cause skin problems) are normally used to de-caffeinate coffee. If you drink decaffeinated coffee, look for organic brands that use a Swiss water-process, carbon dioxide or sparkling-water process (National Geographic,

2008). Health hazards can be avoided by using water or carbon dioxide treated beans instead.

When you consider food options for your family or neighborhood, what can you do *today* to make a difference? What small piece of land could be made productive to increase food availability within walking distance of your home? Take an inventory of what fruit trees are available in your neighborhood (an opportunity to share) and determine which other trees could be planted to increase shade, CO_2 sequestration and provide an additional food supply.

References Cited for Chapter 6
What We Eat

Adams, Mike. (2012). Shock findings in new GMO study: Rats fed lifetime of GM corn grow horrifying tumors, 70% of females die early. *NaturalNews*. Retrieved from http://www.naturalnews.com/037249_GMO_study_cancer_tumors_organ_damage.html

Aquaculture can offset fish deficit. (2013, September 3). *Ghana Business News*. Retrieved from (http://www.ghananewsagency.org/science/aquaculture-can-offset-fish-deficit-scientist--64305)

BALLE. (2012). Retrieved from http://bealocalist.org/

Baudouin, Frédérique. (2013). The most widely used herbicide in the world contains compounds more toxic than declared. *Criigen*, Retrieved from http://www.criigen.org/SiteEn/index.php?option=com_content&task=blogsection&id=5&Itemid=84

Beenen, Isabelle. (2013). Our Food Supply: What You and Your Family Need to Know. *The Sleuth Journal*. Retrived from http://www.thesleuthjournal.com/our-food-supply-what-you-and-your-family-need-to-know/

Benbrook, Charles M. (2012). Impacts of genetically engineered crops on pesticide use in the U.S. – the first sixteen years. *Environmental Sciences Europe*, 24:24 http://www.enveurope.com/content/24/1/24

Benyus, Janine. (1997). *Biomimicry: Innovation Inspired by Nature*, New York, NY: HarperCollins Publishers Inc.

Brown, Paul. (2005). GM crops created superweed, say scientists, Modified rape crosses with wild plant to create tough pesticide-resistant strain. *The Guardian*, Retrieved from http://www.theguardian.com/science/2005/jul/25/gm.food

Burdetter Industries. (n.d.). Global Aquatics. Retrieved from http://www.growfish.com/index.html

Center for Environmental Education. (2008). Retrieved from http://www.ceeonline.org/greenGuide/food/upload/environmenthealth.aspx
City Chicken. (n.d.). Retrieved from http://www.thecitychicken.com/

Craves, Julie. (2006). The problems with sun coffee. Coffee and Conservation: Are Your Beans for the Birds? Retrieved from http://www.coffeehabitat.com/2006/02/the_problems_wi/

Edwards, Andres R. (2010). *Thriving Beyond Sustainability: Pathways to a Resilient Society.* Gabriola Island, BC Canada: New Society Publishers.

Environmental Working Group (EWG). (2013). EWG's 2013 Shopper's Guide to Pesticides in Produce™. Retrieved from http://www.ewg.org/foodnews/summary.php

Fair Trade USA. (2010). What is Fair Trade? Retrieved from http://www.fairtradeusa.org/what-is-fair-trade
Food and Drug Administration (FDA). (n.d.). Natural and Organic Foods. Retrieved from http://www.fda.gov/ohrms/dockets/dockets/06p0094/06p-0094-cp00001-05-Tab-04-Food-Marketing-Institute-vol1.pdf

Gucciardi, Anthony. (2012). Russia Bans Use and Import of Monsanto's GMO Corn Following Study. Nation of Change. Retrieved from http://www.nationofchange.org/russia-bans-use-and-import-monsanto-s-gmo-corn-following-study-1348759383

Gurian-Sherman, Doug. (2008). CAFOs Uncovered: The Untold Costs of Confined Animal Feeding Operations. Union of Concerned Scientists. Retrieved from http://www.ucsusa.org/assets/documents/food_and_agriculture/cafos-uncovered.pdf

Hals, Tom. (2012). Hershey Accused Of Using Cocoa Suppliers That Employ Child Labor. *Huffington Post, Business.* Retrieved from http://www.huffingtonpost.com/2012/11/02/hershey-child-labor_n_2060702.html

ICCAT (International Commission for the Conservation of Atlantic Tunas). (2014). Fisheries and Oceans Canada. Retrieved from http://www.dfo-mpo.gc.ca/international/tuna-thon/iccat-cicta-eng.htm

LabelGMOs.org, (n.d.). We Currently Eat Genetically Engineered Food, But Don't Know It. Retrieved from http://www.labelgmos.org/the_science_genetically_modified_foods_gmo

MacKenzie, Willie. (n.d.). If You Eat Tuna, You Should Know These Five Fish. Greenpeaceblogs. Retrieved from http://greenpeaceblogs.org/2014/05/05/eat-tuna-know-fish/

Mayo Clinic. (2013). Meatless meals: The benefits of eating less meat. Retrieved from http://www.mayoclinic.com/health/meatless-meals/my00752

McDermott, Mat. (2009). *You Wouldn't Eat a Tiger, So Why Would You Eat Endangered Bluefin Tuna?* Retrieved from http://www.treehugger.com/green-food/you-wouldnt-eat-a-tiger-so-why-would-you-eat-endangered-bluefin-tuna.html

McFadden, Steven. (2012). Unraveling the CSA Number Conundrum. *The Call of the Land: An Agrarian Primer for the 21st Century.* Retrieved from http://thecalloftheland.wordpress.com/2012/01/09/ unraveling-the-csa-number-conundrum/

McMenamins. (n.d.). Retrieved from http://www.shopmcmenamins.com/categories/103-Food

National Geographic. (2008) *Green Guide. The Complete Reference for Consuming Wisely.* Washington, DC: National Geographic Society

Northwest Earth Institute (NWEI). (2009). Menu for the Future. Portland, OR.

Ogden, Lillie. (n.d.). The Environmental Impact of a Meat Based Diet, *Vegetarian Times*. from http://www.vegetariantimes.com/article/the-environmental-impact-of-a-meat-based-diet/

Organic Trade Association. (2011). Retrieved from http://www.ota.com/organic/benefits/nutrition.html

Parker-Pope, Tara. (2007, October 22). Five Easy Ways to Go Organic. *New York Times: Health and Science*. Retrieved from http://well.blogs.nytimes.com/2007/10/22/five-easy-ways-to-go-organic/?_r=0

Portlandia Foods. (2010). Retrieved from http://portlandiafoods.com/

Reed, Pamela. (2005). Understanding & Accommodating People with Multiple Chemical Sensitivity in Independent Living. *Independent Living Research Utilization* (ILRU). Retrieved from http://www.ilru.org/html/publications/bookshelf/MCS.html

Robbins, John. (2010). Is there child slavery in your chocolate? *Huff Post Healthy Living*. Retrieved from http://www.huffingtonpost.com/john-robbins/is-there-child-slavery-in_b_737737.html

Rotenberg, L. (2009). Thirteen things you can do to support a sustainable food system. *Menu for the Future*. Portland, OR: Northwest Earth Institute

Savory Institute. (2014). Removing Barriers to Successful Land Management. Retrieved from http://www.savoryinstitute.com/our-work/

Schoepke, Evan. (n.d.). DIY: Urban Aquaculture Manual. Retrieved from http://www.scribd.com/doc/12685513/DIY-Urban-Aquaculture-Manual

Steffen, Alex (editor). (2011). *World Changing: A User's Guide for the 21st Century.* New York, NY: Abrams Books.

USDA. (2012). What is organic? Retrieved from http://www.ams.usda.gov/AMSv1.0/ams.fetchTemplateData.do?template=TemplateC&navID=NationalOrganicProgram&leftNav=NationalOrganicProgram&page=NOPConsumers&description=Consumers&acct=nopgeninfo

Walsh, Bryan. (2013). *The Pacific Bluefin Tuna Is Going, Going.... Time, Science and Space*. Retrieved from http://science.time.com/2013/01/11/the-pacific-bluefin-tuna-is-almost-gone/

Walljasper, Jay. (2007). *The Great Neighborhood Book: A Do-It-Yourself Guide to Placemaking.* Gabriola Island, BC, Canada: New Society Publishers.

Watson, Stephanie. (2012). Organic food no more nutritious than conventionally grown food. *Harvard Health Publication*. Retrieved from http://www.health.harvard.edu/blog/organic-food-no-more-nutritious-than-conventionally-grown-food-201209055264

Wines. (2013, March 15). Fish Populations in the United States Rebound. *New York Times*. Retrieved from http://www.nytimes.com/2013/03/16/science/earth/fish-populations-in-us-rebound-since-1996-catch-limits-law.html?_r=0

Worldwatch. (n.d.). Local Food, Did you know …? Retrieved from http://www.worldwatch.org/system/files/Local%20Food.pdf

Yale-New Haven Hospital. (n.d.). Chocolate: Food of the Gods. Retrieved from http://www.ynhh.org/about-us/chocolate.aspx

CHAPTER 7
SUSTAINABLE AGRICULTURE AND PERMACULTURE

The best way to start this chapter is with a quote from *The Sixth Extinction,* (2001) by Niles Elderedge: "Agriculture represents the single most profound ecological change in the entire 3.5 billion-year history of [human] life."

Major crop production in the past 50 years has more than doubled, and production of cereals has tripled. The amount of food calories potentially available for each person on earth rose 13% in the past 30 years. Higher-yielding, faster-maturing crops are responsible for these increases rather than access to more farmland (Hawken et. al., 1999). When we came to this continent and cut the forests and plowed the prairies, we did not understand what we had, so we really did not know what we were doing or what ecosystems we were undoing. We replaced the native ecosystems with an endless expanse of croplands. Just fly across the country and look out the window. Where do native ecosystems exist? Even our forests are cultivated as a crop.

There are many disturbing facts about our agricultural practices (Hawken et. al., 1999; Benyus, 1997):

- 1% of Americans grow the food for the rest of us;
- 87% of our food comes from only 18% of the farms (Many farms are factories with absentee owners);
- 50% of the farms are operated by 7 companies;
- A 50% increase of food production is due to irrigation; and
- From 1961-1996 nitrogen fertilizer usage increased 645%.

Due to corporate farming, the nutritional value of our food has declined. Thomas (2003) compared and contrasted the chemical constitution of common foods. He studied the important organic and mineral constituents of 27 varieties of vegetables, 17 varieties of fruit, 10 cuts of meat, and some milk and cheese products. The results showed "a significant loss of minerals and trace elements in these foods" over the period of study, 1940 to 1991 (Thomas, 2003). A similar study was done by Davis et al., (2004). Their results also showed declines for calcium, phosphorous, iron (15% decline), riboflavin and ascorbic acid (20% loss), a decline in protein of 6%, and 38% for riboflavin.

Our subsidized Interstate Highway system is what makes it possible to transport our food an average of 1,300 miles (2,092 km) in the United States. What would happen if this transportation system broke down for some reason? Our commercial farms appear to be successful, but many problems are hidden:

- One third of our topsoil has eroded and the rest degraded (soil erosion costs the world 400 billion dollars each year);
- Soil is eroding faster that it can form: growing one bushel (0.04 m^3) of corn on a conventional farm erodes 2 to 5 bushels (0.07 to 0.18 m^3) of topsoil;
- Soil health, the microorganisms, are in decline.

(Hawken et al., 1999)

Healthy soil contains 4,000 distinct organisms per 0.04 ounces (1 g) of soil, depending on location (Hawken et al., 1999). We know very little about most soil organisms: for every 10 microbes we see on plant roots, we can grow only one in the lab (Hawken et al., 1999). Since 1900, productive land per capita has declined from 14 acres to 3.7 (5.7 ha to 1.5 ha), and less than 1 acre (0.4 ha) is arable

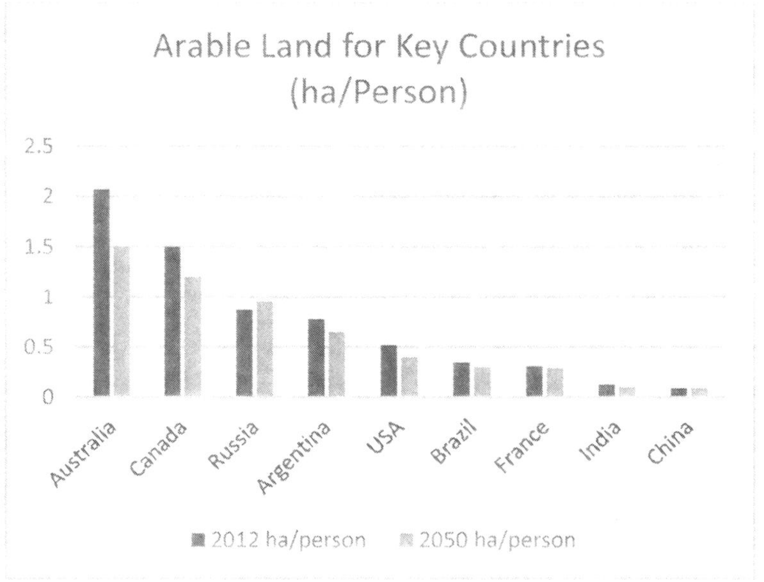

Figure 7.1. Arable land available per person in various countries.
Source: The World Bank Group:
http://data.worldbank.org/indicator/AG.LND.ARBL.HA.PC

(Figure 7.1). Land required to support the average person in the United States has risen from 2.5 acres (1 ha) per person to an average of 10 acres (4 ha) (Hawken et al., 1999). This takes into account population increases and simultaneous soil degradation. Our conventional agricultural system is in contrast to nature's. Our monoculture mentality causes disease to destroy 13% of the world's crops; insects take 15%, and weeds take 12%. Therefore 40% of every harvest is lost in the field (Hawken et al., 1999). In 1948, before we used pesticides, we lost 7% of the crop to insects. Today we use 20% more insecticides; our crop loss to insects is now 13%, AND we are killing all the beneficial insects as well (Hawken et al., 1999). The costs of our agricultural successes have been too great.

However, we do have alternatives. Sustainable choices are already showing their worth. They are more energy efficient, and they use natural biological sources for increasing soil fertility and managing pests (Edwards, 2005). Families are encouraged to stay on their farms and rural communities are enhanced.

According to the Bureau of Labor Statistics, 60% of the farmers in this country are 55 years old or older. The age of the main farm operator has increased from 54 years of age in 1997 to 57 years of age in 2007 (USDA, 2009). The principle farm operators, aged 65 years old or older, have increased 10% since 1969 (USEPA, 2013). This causes concern about the health of the institution of the American family farm (US EPA, 2013). The National Young Farmers Coalition (n.d.) and the Greenhorns (n.d.) are organizations that help new young farmers get trained and start careers in producing food for the country.

If we had more localized food production, transportation could be reduced; we could produce a superior product and at the same time provide much greater diversity (Hawken et al., 1999). Remember the word of the year, "locavore" (Edwards, 2010), as described in the previous chapter. We need to make food a *local* concern.

Current conventional agricultural practices in the U.S. require 45,000 ft^2 (0.42 ha) to feed just one person on a high-meat diet; only 10,000 ft^2 (0.093 ha) are necessary for a vegetarian diet. If the methods of biointensive gardening are used, a diet can be provided for a vegetarian using only 2,000 to 4,000 ft^2 (0.02 to 0.04 ha). To get started, only a piece of land and small locally produced hands tools are needed – no capital, no chemicals (Hawken et al., 1999).

Other practices also show promise. Recently rediscovered author Masanobu Fukuoka (1978) developed a method of farming that simplifies agriculture. He looked at everything that was being done in conventional agriculture and tried to find a way to avoid those actions, creating the "Do-Nothing" system of organic farming. The crops follow nature and look after themselves (Fukuoka, 1978). His system for a one-quarter acre yields 22 bushels (0.78 m^3) each of rice and winter grains (enough to feed 5 to 10 people for a year) with only the work of 1 to 2 people for a few days each year (to sow and to harvest). He spreads waste straw on the fields and grows a crop of clover to add nitrogen to the soil naturally, which creates no pollution. The grains sprout up through the clover. Many crop diseases are caused by adding too much nitrogen to the soil, using too much irrigation water, and by using weak, so-called "improved" seeds. If strong roots are encouraged, chemicals are not necessary and can be avoided (Fukuoka, 1978).

It is important to grow food wherever you can find space. Even in a window box or in containers on a balcony, a significant amount of food can be grown. Our balconies should be burgeoning with fruit and vegetables. Vertical gardens provide an additional option for a garden (Figure 7.2). Discarded wooden pallets can become a small garden by planting between the wooden slats that prevent weeds from growing. An arbor over a walkway provides a unique place for vines to grow, simultaneously increasing garden space, keeping vegetables within easy reach up off the ground and away from pests (Figure 7.3).

Figure 7.2. Two forms of vertical gardens: (Right) Shoe organizer gardens. Source: Growing Veggies in Small Spaces: Lowes Home Improvement. Reprinted with permission; (Left) Soda bottle garden wall. Source: Brazilian Design Studio: Rosenbaum. Reprinted with permission. http://www.treehugger.com/sustainable-product-design/awesome-vertical-garden-with-recycled-pet-bottles-at-poor-family-home-in-sao-paulo.html

Government subsidies conceal the actual costs of growing and transporting our food. The issue is to find a way to educate the public about organic farming, so that consumers understand the "true costs" (Edwards, 2005). There is a massive, global food-transportation network, supported by governmental subsidies. What happens if those subsidies end? Think of all the off-season food that we eat because of this global network: apples from Chile; mussels from New Zealand; frozen catfish from China; green beans from Belgium (Trader Joe's sells beans from France); jalapenos from Peru; and blackberries from Guatemala. Food that is transported thousands of miles to local supermarkets disrupts the local economy because these markets become tied to international market fluctuations (Bridges, 2007; Edwards, 2005).

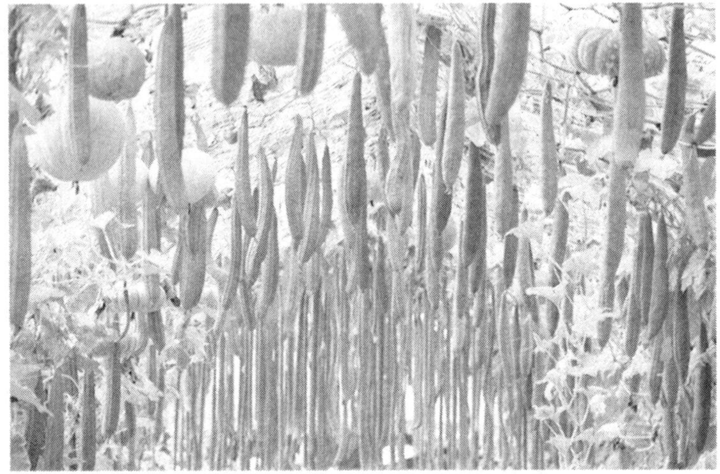

Figure 7.3. Gourds hanging over a walkway provide increased garden space, easy access to vegetables when ripe, and prevents access to pests. Source: Bigstock.com.

A better choice might be what is called permaculture, a term derived from "permanent agriculture." This form of agriculture attempts to copy the actions of a natural ecosystem with a conscious design that includes diversity, stability and resilience (Edwards, 2005). Bill Mollison is a researcher, author, scientist, teacher, and naturalist, and is considered to be the "father of permaculture.". The Mollisonian Permaculture Principles are a model that we can use as an example for an individual sustainability model. As described by Edwards, the principles are as follows:

1. "Work with nature, rather than against the natural elements, forces, pressures, processes, agencies and evolutions, to assist rather than impede natural growth."

2. "The problem is the solution. Everything is a positive resource."
3. "Make the least change for the greatest possible effect. The yield of a system is theoretically unlimited. The only limit is the imagination of the designer."
4. "Everything gardens, or has an effect on its environment." (Edwards, 2005)

In a city, space is frequently very limited, we have access to little or no land, and there can be many rules to follow. The Permaculture Institute (2013) offers solutions for some of these challenging situations. They recommend the development of cooperative arrangements within your neighborhood to ensure long-term success of common gardening projects and shared equipment. If space is limited, grafting multiple varieties onto existing rootstock means a single tree can bear several types of apples (Permaculture, 2013). Find out what trees are in your neighborhood, and plant those that are missing in your own yard. Sharing is a natural way to meet neighbors and to come to trust each other (Hemenway, 2007). Worm composting changes food wastes into odor-free compost for the garden. Become a "Backyard Forester" to increase the canopy in the city, and to restore health to watersheds, ecosystems and neighborhoods. The more you do within your own neighborhood will prevent the conversion of other native land into ecological deserts (monocultures) (Hemenway, 2007). To hold water in the soil, follow nature's example and bury old, rotting wood in a trench; then backfill and plant the surface, for example, with blueberries (Hemenway, 2009).

If you want to become more involved within the city, many organizations train volunteers to care for a variety of ecosystems.

"Citizen Pruner" in New York trains volunteers in tree health and in pruning techniques to help care for street trees (Permaculture, 2013). In Portland, Oregon, Friends of Trees does similar work (Friends of Trees, 2014). Some cities permit the raising of chickens. Chickens enhance the quality of the environment by aerating soil, eating insects, and chickens provide a free source of protein as a bonus (City Chicken, n.d.).

Using permaculture principles means you share with wildlife (Hemenway, 2009). When deer are a "problem," a two-sided hedge can be grown, supplying a food source for the deer on the outside, a thorny core of wild plums to deter the deer from crossing the hedge, and on the inside domestic fruit varieties. This hedge that provides food for humans as well as food for wildlife is called a "fedge" (Hemenway, 2009).

Some cities, such as Portland, Oregon, map the fruit found on public lands. Edible fruit should not be allowed to go to waste as it ripens. Urban Edibles (n.d.) is a volunteer organization that monitors the fruit ripening on public properties, processes the fruit and nuts, and splits the results between the Food Bank and its volunteers (Steffan, 2011).

You may be thinking, "But gardens are so much work!" Why is that? As gardeners, we have a tendency to ignore nature's rules. This generates a lot of tedious work; it leaves no habitat for native or wild species; and requires the use of many poisonous chemicals (Hemenway, 2009). The way nature gardens is in total contrast to the way that we garden. We clear all the plants and expose bare soil; then plant a single species in a large block making an ecosystem with one plant height and root depth (a monoculture). Nature does none of this. Nature does everything to avoid all of those conditions

(Hemenway, 2009). In addition, nature does not till the soil. According to Hemenway (2009), tilling does the following:

1. Destroys weeds and pumps air to microbes that become supercharged.
2. Releases a flood of nutrients for fast growth.
3. Depletes the soil's fertility.
4. Causes disease.
5. Ruins the soil structure.

If we were to create our gardens the way nature does, we would have much less work to do (see Fukuoka, Masanobu, 1978, *The One-Straw Revolution)*. Our lawns of grass and flowers are like a prairie; if you add an occasional tree or shrub they become like a savanna (Hemenway, 2009). But both of these natural ecosystems, prairies and savannas, flourish under specific conditions: low rainfall, heavy animal grazing, and frequent fires. In contrast, in our front yards, we have no animal grazing, and certainly don't allow frequent fires! We irrigate and fertilize our lawns, and this encourages these ecosystems to transition to shrubland and forest (Hemenway, 2009). To prevent these changes, much time and effort is required to hold back the natural ecological progression. No system can function properly when such mixed signals are provided: Grow! Don't grow!

A typical northern forest recycles 98% of its required calcium; the remainder comes from the weathering of rocks (Hemenway, 2009). We are doing well in our cities if we recycle 30%. Our agricultural lands are heavily fertilized and lose 25% to 60% of their calcium each year (Hemenway, 2009). If we are to have a truly sustainable society, we need to learn how to recycle as well as nature does.

A better option might be to think about having ecological gardens. Once an ecologically designed garden is mature, very little upkeep is necessary to maintain it. Biodiversity increases because wildlife habitat is created rather than destroyed. Air, soil, and water quality improves, and humans have less work to do (Hemenway, 2009).

Most of the food we eat anywhere in the world comes from only 20 species, none of them perennials (Benyus, 1997). Of all plants on the earth, 99.9% are perennials. Perennials have life cycles that are longer than 2 years. Annuals complete their entire life cycle in a single season. Runoff from a wheat field (wheat is an annual) is 8 times more than that from a natural prairie (which contains mostly perennials) (Benyus, 1997). Perennials are also self-fertilizing (30% of the roots decay each year, adding organic matter to the soil) and self-weeding. Two-thirds of a perennial's root system survives the winter allowing the plant to emerge earlier than annuals that must start from seed each year (Benyus, 1997).

Wes Jackson, President of the Land Institute, says that after 17 years of research, the Institute is convinced that we <u>can</u> build an agricultural paradigm completely different to the one we have used for the past 8,000 to 10,000 years (Benyus, 1997). Most of our food crops are exotics (originating from another country). We have very few domesticated native plants: sunflowers, cranberries, blueberries, pecans, concord grapes, and Jerusalem artichokes are the only ones (Benyus, 1997).

How does nature build an ecosystem? Prairies are biologically quite diverse, with more than 230 species on a single hill. There is not a single warm-season grass, but 40 types (Benyus, 1997). The prairie is planted with 20 to 30 nitrogen-fixing legumes, not just one. According to Benyus (1997), the species composition in the prairie remains unchanged, but in each year a different species excels.

Because of this variation, pests have a harder time finding their targets.

If nature likes 230 species, exactly how many are required for us to produce the same overall effect? Surprisingly, it may be as few as 8 (Benyus, 1997)! Is there a perennial that can produce as much seed as wheat? In an initial experiment, and without breeding to increase yield, 2 perennials have already reached the same yields (800 pounds per acre or 907 kg per ha): Illinois Bundleflower and Wild Sienna. Two steps are needed to make the transition to these alternative perennial crops: new recipes and the development of a taste for the new crops (Benyus, 1997).

It is well documented that our agricultural soil has been degraded. What is the best way to enrich the soil naturally? A system referred to as composting, which uses 30 parts carbon to 1 part nitrogen, is the best. According to Hemenway, (2009) organic material that is high in carbon is "brown" (such as dried leaves, hay straw, wood shavings); "green" materials are high in nitrogen (grass clippings, plant trimmings, kitchen waste, manure are examples). When you mulch your garden, how much mulch do you normally apply? A couple of inches (maybe 5 cm)? Think again of what may be appropriate (Figure 7.5). The best way to enrich soil is to sheet mulch in the fall. To sheet mulch an area requires the layering of a biodegradable sun block, usually cardboard, covered by at least 8 to 10 inches (20 to 25 cm) of organic matter, both green and brown materials (Hemenway, 2009).

Hemenway (2009) writes: "One fall I was extending my garden into a struggling lawn. I decided to use sheet mulch for the new beds. By spring, my lifeless red clay soil had darkened to chocolate brown, was seething with worms, and had begun to fluff to a marvelous

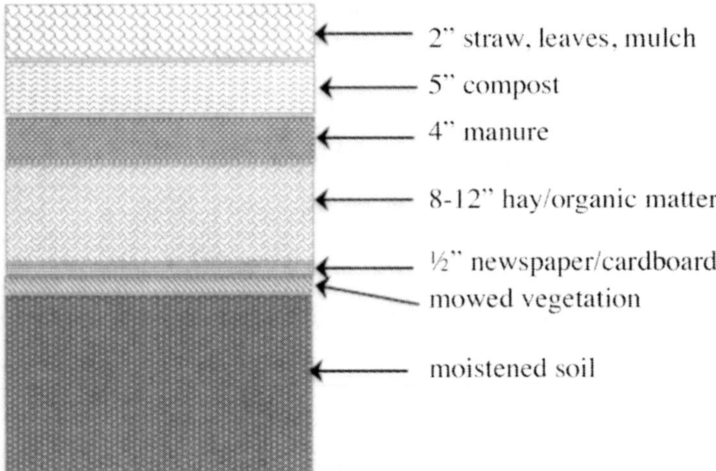

Figure 7.5. Sheet mulching. Source: Prock, 2010.

crumbly tilth" (Hemenway, 2009). The first requirement is to lay down a weed barrier of cardboard or paper. Add a foot of moist, decomposable organic matter. In the spring you will have great soil. With a little research, you'll find many of these items available locally for free:

- Lumber mills offer bark and sawdust;
- At the coast there is plenty of seaweed and salt hay;
- Agricultural areas have straw, grain hulls, and food industry by-products;
- Canneries and food warehouses have organic wastes;
- Utility companies and landscapers have chipped tree trimmings; and
- Our neighborhoods have plenty of leaves in the fall (collect only those from lawns, not from the street where they may contain fossil fuel residues) (Hemenway, 2009).

Things NOT to add include pet droppings, anything treated with herbicides or pesticides, and meat (especially if the mulched area is uncovered) (Gabriel, 2011).

Another practice, one that is 2,000 years old, called Biochar, uses a mixture of charcoal, bone, and manure. Agricultural waste is converted into a soil enhancer by adding the Biochar mixture to relatively infertile soil. The primary function of Biochar is carbon sequestration; it is used to improve water quality, increase soil fertility, and raise agricultural productivity (International Biochar, 2013). It is relatively inexpensive, widely applicable, quickly scalable (it has been tried at a small scale and can quickly be duplicated at a larger scale), and, if implemented world-wide, could store 2.2 gigatons (2.2 billion tons or 2 billion tonnes) of carbon in the earth's soil annually by 2050 (International Biochar, 2013).

We need to consider the ramifications of all our choices. People who consider themselves to be sustainability-minded tend to choose cotton clothes over those made of synthetics. But cotton is the world's "dirtiest" crop, because it uses the pesticides most hazardous to humans and animals (Organic Trade, 2013). Although land covered by cotton crops represents only 2.5% of the world's agricultural land, cotton uses 16% of the world's insecticides, more than any other major crop (Organic Trade, 2013). Are there any fiber options to replace cotton? There are! We already have at least three options:

1. Bamboo: It has a stronger per unit-mass than steel.
2. Kenaf: An East Indian hibiscus similar to okra and cotton, it may be a possible wood substitute, yielding several times as much fiber per acre as wood.

3. Hemp: It yields 20 to 30 tons of fiber per acre (45 to 68 t per ha), exceeding the amount of fiber acquired for paper from most trees.

(Hawken et al., 1999).

Agricultural residues are often wasted, burned, rotted, or landfilled. That waste totals 280 million tons (254 million t) a year (in 1994), equal to the paper consumption of the entire world (or equal to the total U.S. wood harvest annually). There are alternatives to this kind of waste. For instance, in Nepal, making paper from banana stems is a new sustainable practice. After the bananas are harvested, the plant stems are cut and boiled to make paper pulp. This provides jobs for many local people (Banana Paper, n.d.). In China, Styrofoam packaging "peanuts" litter the landscape everywhere (McDonough and Braungart, 2002). Rice stalks could be used instead of styrofoam, maybe with a few native plant seeds included, and if discarded, would quickly decompose and enrich the environment (McDonough and Braungart, 2002). In Oregon, farmers used to burn the straw after the wheat harvest. Now they leave 90% in the field to prevent soil erosion and to improve soil quality from the straw's decomposition; the farmers sell the other 10% to a Canadian firm to make chlorine-free paper. This small change increases farm income 25% to 50% (Hawken et al., 1999).

Agriculture may take on a totally different shape in the future. A Master's program in landscape architecture at the University of Amsterdam documents the projects of their students. The students give us a glimpse into the future of urban farming and illustrate the scale that may be required to feed our cities locally, with massive grids of agricultural plots hanging over our cities, perhaps like the hanging gardens of Babylon (Placella, n.d.). To be efficient, green is going to have to take on a much larger urban scale.

Finally, learn to "read" nature. If you want to grow good pasture, start with a field of alfalfa and add 4 species of grass. Other wild plants will begin to appear naturally. When you hear a strange crackling noise in the field, it will be the sound of a healthy pasture, as hundreds of thousands of earthworm holes open up after a rainstorm (Benyus, 1997). It may take 3 years more before the birds return, but then you can assess the health of the area by the diversity of birds present. "Learn to read nature instead of relying on the pesticide salesman" (Benyus, 1997).

We *will* have to re-engage farming at some scale post peak-oil, when fossil fuels become too expensive to extract from the ground. It is strange that Cuba, a country that we have mistreated for decades, may provide us with guidance. Due to the dissolution of the Soviet Union in 1989, Cuba experienced a prolonged period of economic hardship from a U.S. trade embargo placed on the island. There were severe shortages of anything related to fossil fuels, such as gasoline and chemical fertilizers. Cubans were forced to restructure every aspect of their lives. In the documentary "Cuba: the Power of Community," we are shown that it is possible to thrive without fossil fuels, and it gives us hope (Cuba, 2011). If we are fortunate, we will learn by their example before we are forced to do so, when fossil fuels are no longer available for us. The Cubans are a resilient people; they call their difficult transition "The Special Period" (Cuba, 2011). What might we call a similar transition? The last time we had a difficult transition period, we called it "the Great Depression."

References for Chapter 7
Sustainable Agriculture and Permaculture

Banana Paper Making. (n.d.). Retrieved from http://nepal.helvetas.org/en/country_programme/innovation_new/banana/

Benyus, Janine. (1997). *Biomimicry: Innovation Inspired by Nature*. New York, NY: HarperCollins Publishers Inc.

Bridges, Andrew. (2007). *Imported Foods Rarely Inspected*, USA Today, Retrieved from http://usatoday30.usatoday.com/news/nation/2007-04-16-imported-food_N.htm

City Chicken. (n.d.). Retrieved from http://www.thecitychicken.com/

Cuba: the Power of Community. (2011). Retrieved from http://www.youtube.com/watch?v=L2TzvnRo6_c&feature=related

Davis, Donald R., Melvin D. Epp, and Hugh D. Riordan. (2004). Changes in USDA Food Composition Data for 43 Garden Crops, 1950 to 1999. *Journal of the American College of Nutrition*, Vol. 23, No. 6, 669–682.

Edwards, Andres. (2005). *The Sustainability Revolution: Portrait of a Paradigm Shift.* Gabriola Island, BC, Canada: New Society Publishers.

Eldredge, Niles. (2001). *The Sixth Extinction.* Action Bioscience. Retrieved from http://www.actionbioscience.org/evolution/eldredge2.html#primer

Friends of Trees. (2014). Retrieved from http://www.friendsoftrees.org/

Fukuoka, Masanobu. (1978). *The One-Straw Revolution.* New York, NY: New York Review Books.

Gardens by Gabriel. (2011). Compost Part II: Let it rot. Let it roll! Retrieved from: http://www.gardensbygabriel.com/blog/category/conservation-tip/feed/

Greenhorns. (n.d.). Retrieved from http://www.thegreenhorns.net/category/about/aboutus/

Hawken, Paul, Amory Lovins, and L. Hunter Lovins. (1999). *Natural Capitalism*. New York, NY: Little Brown and Company.

Hemenway, Toby. (2009). *Gaia's Garden: A Guide to Home-Scale Permaculture*. Second Edition. White River Junction, VT: Chelsea Green Publishing.

International Biochar Initiative. (2013). Environmental Benefits of Biochar. Retrieved from http://www.biochar-international.org/biochar/benefits

McDonough, William and Michael Braungart. (2002). *Cradle to Cradle: Remaking the Way We Make Things*. New York, NY: North Point Press.

National Young Farmers Coalition. (n.d.). Retrieved from http://www.youngfarmers.org/about/our-work/

Organic Trade Association. (2013). Cotton and the Environment. Retrieved from http://www.ota.com/organic/environment/cotton_environment.html

Permaculture Institute. (2013). Retrieved from http://www.permaculture.org/

Placella, Nicola. (n.d.). *The Hanging Gardens of Barcelona*. The Green Dream. Retrieved from http://www.thewhyfactory.com/?page=project&project=18&type=future

Prock, Donna. (2010). *Thinking Forward, Starting Now: Growing your Permaculture Garden*. Retrieved from http://www.cedarmill.org/news/410/growing-permaculture.html

Thomas, David. (2003). A study on the mineral depletion of the foods available to us as a nation over the period 1940 to 1991. *Nutr Health*. 17(2):85-115. Retrieved from http://www.ncbi.nlm.nih.gov/pubmed/14653505

Turtlewoman. (2013). *Difference Between Organic Eggs, Pastured Eggs, And Free Range*. Retrieved from http://hubpages.com/hub/How-to-buy-the-healthiest-eggs#

Urban Edibles. (n.d.). A Community Database of wild food sources in Portland, OR. Retrieved from http://urbanedibles.org/search

US USDA. (2009). NASS. *2007 Census of Agriculture*. 10 Dec. 2009. Retrieved from http://www.agcensus.usda.gov/Publications/2007/Full_Report/usv1.pdf

US EPA. (2013). Ag 101, Demographics. Retrieved from http://www.epa.gov/agriculture/ag101/demographics.html

CHAPTER 8
THE BUILT WORLD

Globally, buildings use 40% of all raw materials (Edwards, 2005). Many builders use the minimum acceptable conditions: they "meet code." Do you think building to minimum standards is going to get us where we want to be? Often regulations and zoning laws prevent us from building the sustainable future we want or envision for ourselves.

What causes inefficiency in our current building practices? Standard practices according to Hawken et al. (1999) include the following sequence of events:

- Take a previously successful set of drawings;
- Change the name on the project;
- Submit the drawings to the client;
- Construct the building;
- Client complains about discomfort;
- Wait for the client to quit complaining; and
- Repeat the process.

We have to change the way we think. We have to avoid following existing paths, because they only take us where we have already gone. A new path needs to be forged.

Regarding comfort, few people ever experience real comfort – thermal, visual, or acoustic (Hawken et al., 1999). Think about what we are getting versus what we need. The part that is missing is quality. Both money and energy can be saved if we begin to design intelligently. If we add thick insulation to our homes and high-quality windows, we can eliminate the need of a furnace. If we use better appliances, little or no waste heat is generated and we can eliminate the need for an air conditioner! And it may not be expensive. Super-efficient houses can cost less than the original, unimproved versions (Hawken et al., 1999).

One solution to the loss of heat and cooling is use of 3 to 4 layers of glazing, special coatings, and low-conductivity gases (such as krypton) in super-high-performance windows (superwindows) (Roberts, 2011). Superwindows, a term coined by Dariush Arasteh, a scientist at Lawrence Berkeley National Laboratory, could save up to half of all U.S. electricity currently used (Roberts, 2011; Hawken et al., 1999) (Figure 8.1). Although they cost 10-15% more than double-glazed windows, superwindows keep us warm in the winter and cool in the summer because they insulate 4 to 5 times better (Hawken et al., 1999). Buildings can be kept comfortable in temperature ranges from -47°F to 115°F (-43.9°C to 46°C) without any heating or cooling equipment! Even better insulating windows being developed: Aspen Aerogel insulates several times better than even superwindows (Hawken et al., 1999).

An organization that aims to make builders aware of the impact of their choices is the U.S. Green Building Council and their green building rating system, the Leadership in Energy and Environmental Design, LEED (Edwards, 2005; LEED, 2013). The council works to

Figure 8.1. Heat lost from non-insulated buildings (red/dark) versus super-insulated ones (green/light). Source: Bigstock.com.

transform how we think about our buildings and communities (LEED, 2013), and provides third-party site evaluations of water and energy efficiency, materials and resources, and indoor environmental quality (Edwards, 2005). Three certification levels are offered: Silver, Gold and the highest level, Platinum. Certification guidelines can also be downloaded for neighborhood development (LEED-ND) so that neighborhoods also have a framework from which to begin change (Steffan, 2011).

The Living Building Challenge, a building certification program of the International Living Building Institute, represents the highest level in building construction (Edwards, 2010; Steffan, 2011). Since its goal is to copy nature and life systems, these buildings generate their own energy and process their own wastes. The challenge has seven performance areas, or "petals." The petals include: site, water, energy, health, materials, equity, and beauty (Living Building, n.d.). The Living Building Challenge requirements are by far the most difficult to accomplish, and use the most advanced measures of sustainability (Edwards, 2010). There are more than 90 sites currently trying to achieve certification, but only 5 have reached this goal. Of those working toward certification, the June Key Delta Community Center in Portland is one of the few sites aiming for this certification in the state of Oregon (ILFI, 2010) (Figure 8.2).

Figure 8.2. June Key Delta Community Center (JKDCC), Portland, Oregon. Transformed from a gas station to a local community center, it is made mostly from recycled local materials, and is striving for the Living Building Challenge certification. Source: Linda Pope.

Why should we design green buildings? Construction costs are decreased; buildings sell and lease faster; buildings are easier to build, operate, and convert to the next use; mechanical systems are smaller and better designed, or eliminated all together (Hawken et al., 1999).

How realistic is it for big buildings to become green, as a retrofit? Lockheed Missiles and Space Company Inc. (now Lockheed Martin) moved to a new location in 1983, Building 157 in Sunnyvale, California. Architects upgraded its windows to reduce its lighting energy requirements by 75% with an expected investment payback in 4 years. What was not included in the projected payback time was the 15% unexpected drop in absenteeism and the simultaneous 15% increase in productivity, all of which paid for the new design in the first year (Hawken et al., 1999). This lower overhead gave the company a profitability advantage permitting it to receive an unexpected contract: the profits from this contract gave Lockheed more than they had paid for the whole building (Hawken et al., 1999). Also in California, an old concrete warehouse with few windows was retrofitted with windows providing natural light, and a better air filtration system. Nontoxic materials were used, and energy efficiency was improved. The retrofitting costs were expected to be recaptured in energy savings within 7.5 years. A 45% decrease in employee absenteeism was an unexpected bonus (Hawken et al., 1999).

One of the most energy-efficient buildings in Europe is a bank built in Amsterdam in 1987. The bank stipulated at the onset that the new construction methods would cost nothing extra (Hawken et al., 1999). A rainwater capture system would irrigate the gardens, both inside and out; offices would have natural air and light; heating and cooling would be as passive as possible. The 92% energy savings paid for the installation of the energy saving systems in only three months. The interior spaces are so appealing that the bank workers hold social and cultural events in the bank (Hawken et al., 1999). When was the last time you had a party at *your* bank?

In Santa Fe, New Mexico, an ugly steel and glass box (originally built as a jail in the 1960s) was transformed in 1991 to an adobe-style inn that looks hundreds of years old. Although the inn was expensive, it broke even in its second year (something rare for new hotels). Hawken et al. (1999) describe its gourmet restaurant's practices as using 90% local, organic ingredients; leftovers go to homeless shelters; kitchen scraps go to a pig farm; and table scraps are composted. This should be the model used by all restaurants.

Rocky Mountain Institute (RMI), originally housed in the Boulder, Colorado, home of chief scientist Amory Lovins, was designed to eliminate both heating and cooling systems. The 3,850 ft^2 (358 m^2) home cost $500,000 to build ($130 per ft^2 or $1,397 per m^2), including the land and many built-in features, equal to the market rate for homes in 1984 (when it was built) around Aspen, Colorado. An additional $6,000 was spent for the energy and water efficiency technologies, but this saved $19 per day, and the extra expenditure was recouped in only 10 months (RMI Residence, 2013). The inner gardens include banana trees that have produced bananas without the use of a furnace (Hawken et al., 1999), even with average low temperatures of 18°F (-7.7°C) in January.

The worldwide average energy use is 2,000 watts per person per year (Steffan, 2011). However, we do not all use energy at the same rate. Western Europe has an average of 6,000 watts per person per year; the United States has an average of 12,000 watts per person per year; the energy use for the Global South is negligible (Novatlantis, 2007) (Figure 8.3). To have a sustainable world, we have to uplift everyone. That means if the Global South is to reach the same standard of living that we have in the developed countries, the developed countries have to use less. How can we do this and not feel that we are doing without? We have to choose better materials, change our infrastructure, and change our personal choices, so that living in a 2,000-watt society leaves no one in the dark (Steffan, 2011). This goal of using only 2,000 watts per person per year and being carbon-neutral does not mean that our quality of life will suffer. In actuality, our quality of life will improve: we will get rid of waste and excess (Steffan, 2011). Several cities, including Zurich,

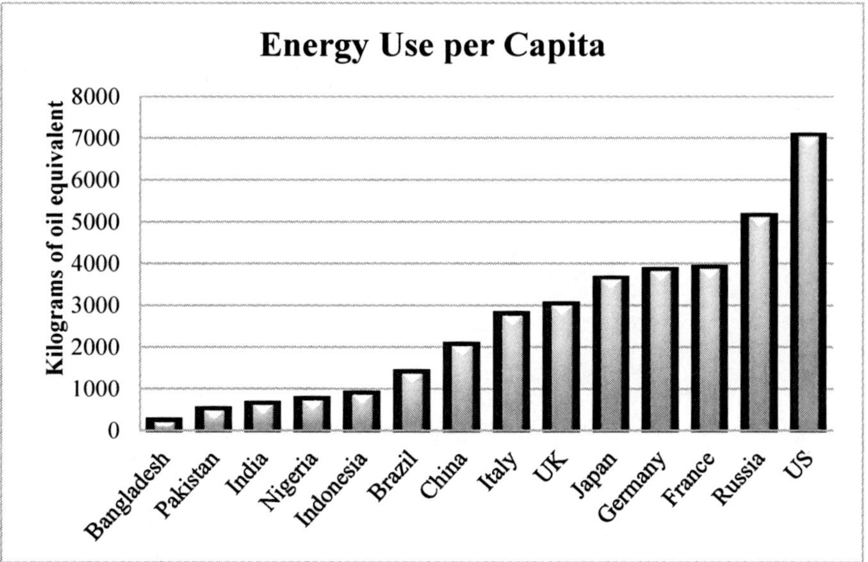

Figure 8.3. Amount of energy required in various countries. Data from the World Bank. Source: Wikipedia

Switzerland, are well on their way to becoming a 2,000-watt society. Through efficient technologies, renewable energy, and a more economical handling of material resources, Zurich has set the goal to go from 5,000 to 2,000 watts of energy-per-person by 2050. They also plan to reduce their carbon dioxide emissions from 5.5 metric tons per year to 1 metric ton, also by 2050 (Steffan, 2011).

In much of Europe, something called the Passive House is being built. Scandinavian and German-speaking countries are at the forefront; around the world, 15,000 passive houses have been built (Edwards, 2010; Steffan, 2011). One Passive House in Austria needs almost no heating: body heat, the heat from cooking, and passive solar heat are all that are required; there is no furnace (Edwards, 2010; Steffan, 2011). And yet very few passive houses have found their way to the United States. Why is the U.S. usually 30 years behind European sustainability practices? When will we become the great innovators again?

Växjö, Sweden has set out to become one of Europe's greenest cities. It plans to reduce its emissions 55% by 2015, and 70% by 2025 (Edwards, 2010). In comparison, in 2012, Portland, Oregon, one of the greenest cities in the United States, just announced that it has reduced emissions by 26% (Giegerich, 2012). In 2000-2002, Britain built its first carbon-neutral community: Beddington Zero Energy Development (BedZED) (Edwards, 2010). The community is built on reclaimed land and boasts passive solar energy, rooftop gardens, natural light, renewable energy, and wastewater recycling (Edwards, 2010).

Masdar City in Abu Dhabi has the goal of being the world's first zero-carbon, zero-waste, car-free city (Edwards, 2010). Would we do better if our city, wherever we live, had some outrageous goal? Ray Anderson, CEO of Interface Inc. would respond with a resounding *yes*! All you have to do is set an outrageous goal and people will try (Arratia, 2010)! What outrageous goal can you set in your household or city?

In Britain, in order to reduce energy consumption, neighborhoods compete against each other to see who can make the greatest reduction. People work harder to reduce consumption when they realize that their neighbors are somehow conserving more than they are (Steffan, 2011).

We are making progress in the United States. For example, in Seattle, Washington, the design firm Weber Thompson (2008) designed an Eco-Laboratory that incorporated in its design a rainwater collection system, a hydroponic garden to grow vegetables for the community, and a wastewater treatment system to convert blackwater to greywater and then potable water. Fresh air finds its way into the building through underground "earth tubes;" wind turbines and solar panels provide energy, as do fuel cells powered by the methane by-product of the wastewater treatment system. In all, it uses 30% less energy by doing much of which we have already known for hundreds of years (Thompson, 2008).

In the 1980s, New Urbanism became the first organization to recognize the importance of neighborhoods designed for both pedestrians and cars. The organizations' philosophy includes diverse neighborhoods that:

> promote the creation and restoration of...walkable, compact, vibrant, mixed-use communities composed of the same components as conventional development, but assembled in a more integrated fashion, in the form of complete communities. These contain housing, work places, shops, entertainment, schools, parks, and civic facilities essential to the daily lives of the residents, all within easy walking distance of each other (New Urbanism, n.d.).

By following New Urbanism's guidelines, communities can become unique, special places where we all love to live. Why go anywhere else when "home" is so great?

Another key organization has also stepped up to the sustainability plate and moved to the forefront in building sustainable communities: The American Society of Landscape Architects (ASLA). Much information is available at their website (http://asla.org). They describe many case studies, including a local parking lot at Mount Tabor Middle School, in Portland, Oregon, and the transformation process of changing the parking lot into a rain garden (Figure 8.4). The rain garden provides a natural way to manage stormwater runoff, reduces radiant heat from the asphalt parking lot, creates wildlife habitat, and provides the opportunity of a learning garden for students. The ASLA's website has several animations describing the benefits of this type of transformation:

- Building a Park Out of Waste
- Designing Neighborhoods for People and Wildlife
- The Edible City
- Energy Efficient Home Landscapes
- Designing for Active Living
- From Industrial Wasteland to Community Park
- Infrastructure for All
- Leveraging the Landscape to Manage Water
- Revitalizing Communities with Parks
- Urban Forests = Cleaner, Cooler Air

(ASLA, n.d.)

Many of the ASLA topics covered on their website are the same topics covered in this book, providing another source of information.

Figure 8.4. Mount Tabor Middle School's (left) courtyard and (right) parking lot, transformed into a rain garden to handle wastewater treatment and increase wildlife habitat. Source: L. Pope.

At ASLA's Sustainable Urban Development webpage, hundreds of resources are offered to help fight urban sprawl, to change to sustainable zoning, to reuse brownfields, to change city centers, and to create open spaces.

If we want better buildings and better communities, what can we do? Start by going to your neighborhood association meetings. Find out what is happening in your own neighborhood. If we come together as communities, we can find the best way to solve problems together. What can help guide this process? One way is in a "charrette." A charrette brings together architects, engineers, landscapers, hydrologists, artists, builders, occupants, commissioners, maintenance staff, etc., so both professionals and community leaders can have a platform that enables them to work together (NCI, 2011). The National Charette Institute trains individuals who aspire to become community leaders (NCI, 2011).

Another resource worth investigating is the Green Infrastructure Digest (n.d.), which describes the many organizations in the Portland, Oregon area that are involved in changing the way our neighborhoods deal with infrastructure and water management.

If there ever was a town that needed help, it was Greensburg, Kansas on May 4, 2007, when almost every building was destroyed by an F5 tornado. Within days, the townspeople came together and decided to rebuild, and to make their town as green as possible. As a result, Greenburg, Kansas, is the greenest town in the United States (Edwards, 2010), with the most LEED certified buildings per capita in the world. Their slogan is "See how we put the 'Green' in Greensburg" (Greensburg, 2013).

Looking at construction details, each new product should address many sustainability concerns, at every scale. For example, the traditional black asphalt shingles that cover roofs absorb heat from the sun's rays and keep homes warm during low temperatures. However, we don't want our homes warm in the summer. During that season, it would be best to have white tiles that would reflect the heat to keep our homes cool. We can't have different tiles for both seasons. Or can we? A company called Thermeleon makes a tile that is black, but as temperatures increase they turn white to reflect solar energy, reducing the energy needed to cool the home (Steffen, 2011). This is a new temperature-sensitive product that solves this particular heating and cooling issue.

Take a minute and think about the cities you like to visit and why you like them. I would wager that they are not cities that require you to drive to everything you want to see or do. Instead, everything is within an easy walk. "Deep walkability" is essential in our communities: "If you can't walk out of a neighborhood, it's not sustainable, no matter how many solar panels and green roofs it has" (Steffen, 2011). What Steffen means is that when you step outside your front door, you should have several options available, and each should result in a pleasant walk. This quality of walkability is something only the people in a community can build. If there are many good places to "hang out," then the street becomes an extension of your home (Steffen, 2011; Walljasper, 2007).

Richard Register (1987) had a great vision for his city of Berkeley, California. He envisioned a 50 to 150-year transformation of the core of his city: neighborhoods would become more dense with more high-rise buildings, but the spaces between the neighborhoods (sprawl) would be replaced with forests and farms (Sterner, 2013). His book is only now being recognized and quoted by others. It is too bad that the value of his visionary plan was not appreciated when he introduced it (Figure 8.5).

The most important thing we can do right now is develop a greater sense of community as we embrace sustainability to its fullest extent. Three books will greatly aid anyone to embrace this challenge: *The Natural Step for Communities: How Cities and Towns can Change to Sustainable Practices*, by James and Lahti (2004);

The Great Neighborhood Book: A Do It Yourself Guide to Placemaking, by Jay Walljasper, (2007); and *Toward Sustainable Communities: Resources for Citizens and Their Governments,* by Mark Roseland (2005). Each author provides inspiring examples for community transformation, and how to lead or participate.

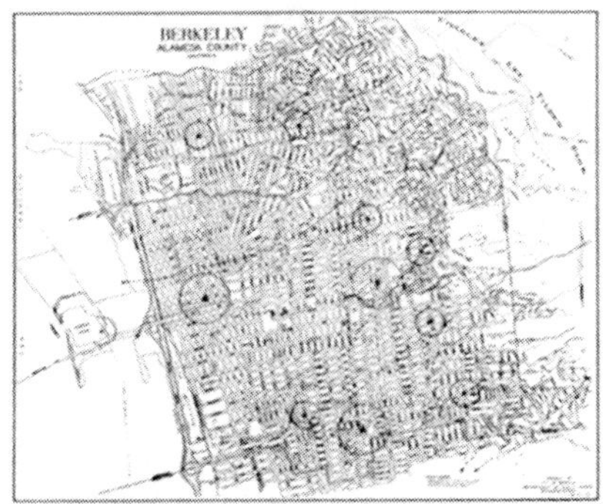

A. Berkeley: 1987

B. Berkeley: with the start of denser centers (in 15 to 50 years)

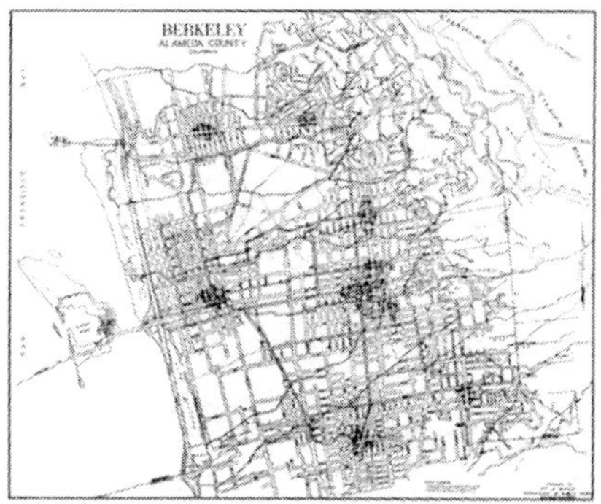

C. Berkeley: in 25 to 90 years

D. Berkeley: in 40 to 125 years with new, denser community centers.

Figure 8.5. Richard Register's vision of decentralization of Berkeley, California, projecting a shift in growth patterns over time, creating better neighborhoods with denser population centers, and including natural ecosystems. Public transportation connects the new neighborhood centers. Source: Richard Register, 1987. Ecocity Berkeley, p.p. 124-129.

One Planet Living has ten guiding principles to direct the change process for everyone in communities.

1. *Zero Carbon* to make buildings more energy efficient and deliver all energy with renewable technologies.
2. *Zero waste* to reduce waste, reusing where possible, and ultimately send zero waste to the landfill.
3. *Sustainable transport* to encourage low-carbon modes of transport, reducing emissions and the need to travel;
4. *Sustainable materials* the use sustainable healthy products, with low embodied energy, sourced locally, made from renewable or waste resources;
5. *Local and sustainable food* that use low impact, local, seasonal and organic diets and reduce food waste;
6. *Sustainable water* that uses water more efficiently in buildings and in the products we buy; and tackling local flooding and watercourse pollution;
7. *Land use and wildlife* to protect and restore biodiversity and natural habitats through appropriate land use and integration into the built environment;
8. *Culture and Community* to revive local identity and wisdom; to support and participate in the arts;
9. *Equity and local economy* to creating bioregional economies that support fair employment, inclusive communities and international fair trade;
10. *Health and happiness* to encourage active, sociable, meaningful lives to promote good health and well-being.

(One Planet Living, 2013)

When each individual, each building, each community is working toward these guidelines, we will attain our sustainable future. But it will take everyone's acceptance of the need to change.

Another organization that is working to implement these guidelines is EcoDistricts (2013). They describe an EcoDistrict as "a new model of public-private partnership that emphasizes innovation and deployment of district-scale best practices to create the neighborhoods of the future – resilient, vibrant, resource-efficient and just" (2013). Their summits held each fall are inspiring events that bring people together from all over the world, to figure out just what it means to be a sustainable city now and in the future, and to develop plans to implement the transformation.

When thinking about communities, the best building materials are of course local. What "local" means will vary, depending upon resources common to that location. Forgotten, and now being rediscovered, local resources have lower carbon footprints due to decreased miles for transport. Local materials also tend to be energy efficient. Two examples are straw-bale homes and rammed-earth homes. The materials as well as the homes are nontoxic, safe, durable, and versatile.

As the world's population increases, we will have the issue of where to put the expected 50% increase in population by 2050. Where in your neighborhood would you put 50% more people? Or should we take over more of our remaining agricultural land, or clear more of our forests to make room for houses? Maybe the place to start would be in rethinking how much space we need to live.

Across the country, a Tiny House movement is growing as people discover that having what they need and no more, frees up both time and resources for other activities (Figure 8.6). Jay Shaffer of Tumbleweed Tiny House Company is something of a hero to people of the Tiny House movement, a small but growing group

Figure 8.6. The Tiny House movement includes a lot of variety, from rural settings to urban ones, rustic to Floridian. Source: (Upper left) iStock; (Upper right) Linda Pope; (Lower left) Reprinted courtesy of Tumbleweed; (Lower right) Shutterstock.com.

"who drastically shrink their living space in hopes of living a cheaper, less wasteful, and happier life" (Leader, 2012). Shafer was attracted to the idea of living with fewer possessions. "When you live in a tiny house you only have room for the things that truly matter," he says about his current 106-square-foot (9.8 m^2) home in Sebastopol, California (Leader, 2012).

Many others have gone even smaller. Designer Kevin Cyr (n.d.) is obsessed with portable housing and mobile living, and he creates portable homes that mounted on bicycles. Alison Futuro (2012) describes beautiful designs for bicycle homes being built in China. These designs are clean, simple, and elegant. With so many homeless and/or very mobile citizens around the world, these small homes may be a solution.

"You have to choose what's essential," says Shaffer. He also says that once you live in a tiny house, you find you no longer need anything larger (Leader, 2012). Instead of thinking, "What is the biggest house we can afford?" perhaps we need to think instead, "How small can I go and still have a high quality of life?" There are many sites on the internet that offer Tiny Houses for sale. The reasons some owners give for why they are selling their tiny houses are sometimes shocking. "I've decided this tiny house is too big. I can go smaller!" The current challenge for Tiny House owners is to work within the current zoning laws to allow this new form of home. Most building codes do not allow buildings under 400 ft^2 (37 m^2). By being constructed on wheels, on a trailer, the tiny house gets around the issues of code. Tiny houses are usually between 96 and 200 ft^2 (8.9 m^2 to 18.6 m^2).

In order to consider more fully the impact of 50% more people in your community, compare the following two maps of Portland, Oregon: one from 1806 as settlers were beginning to arrive at the confluence of the Columbia River and the Willamette River (Figure 8.7), and the second map from 200 years later, in 2005 (Figure 8.8). What will our city be like with 50% more inhabitants? Will the urban footprint be more dense or will we lose more forests and farms to urban sprawl?

Look for new ways to change your neighborhood to allow for growth from within: basement apartments, attic dwellings, apartments over a garage, accessory dwelling units. Or provide spaces where a tiny house can find a home. How can your community be the best that it can be? What does your neighborhood need to be an amazing place? Each person in the community has the responsibility to make this transformation occur. Each community will be unique in the approach it uses to accomplish this new direction.

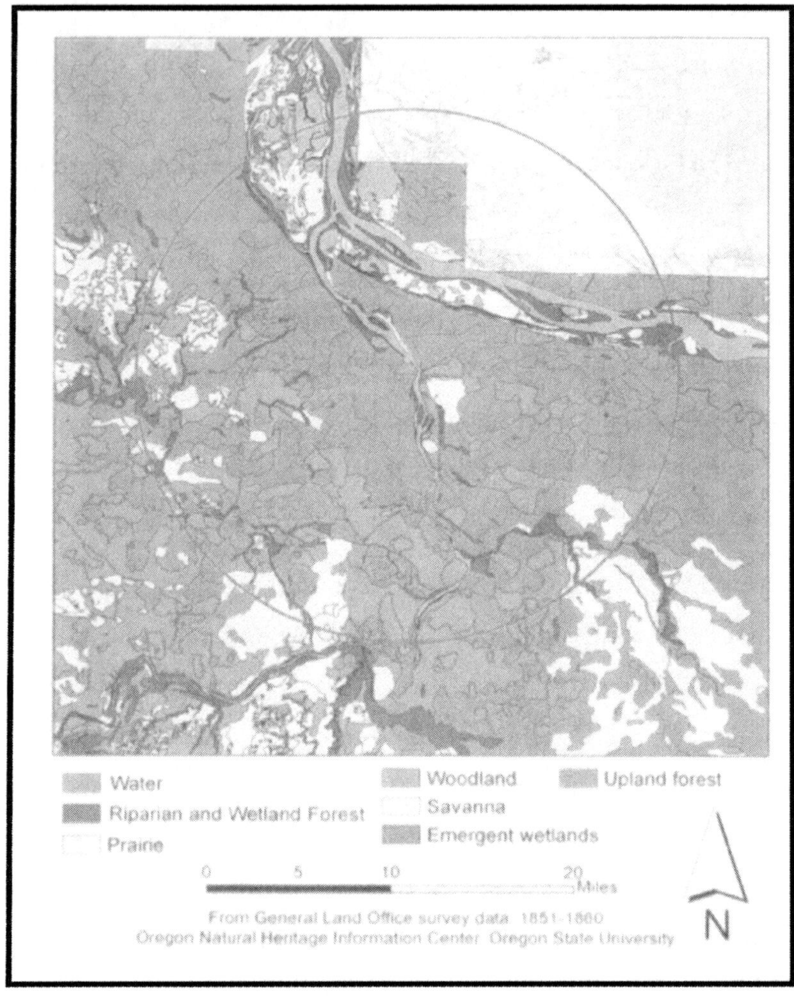

Figure 8.7. Portland, Oregon, circa 1806 when the landscape was completely natural, all forests and wetlands. Source: Christy, J.A., A. Kimpo, V. Marttala, P.K. Gaddis & N.L. Christy. 2009. Urbanizing flora of Portland, Oregon, 1806-2008. *Native Plant Society of Oregon Occasional Paper* 3: p. 33

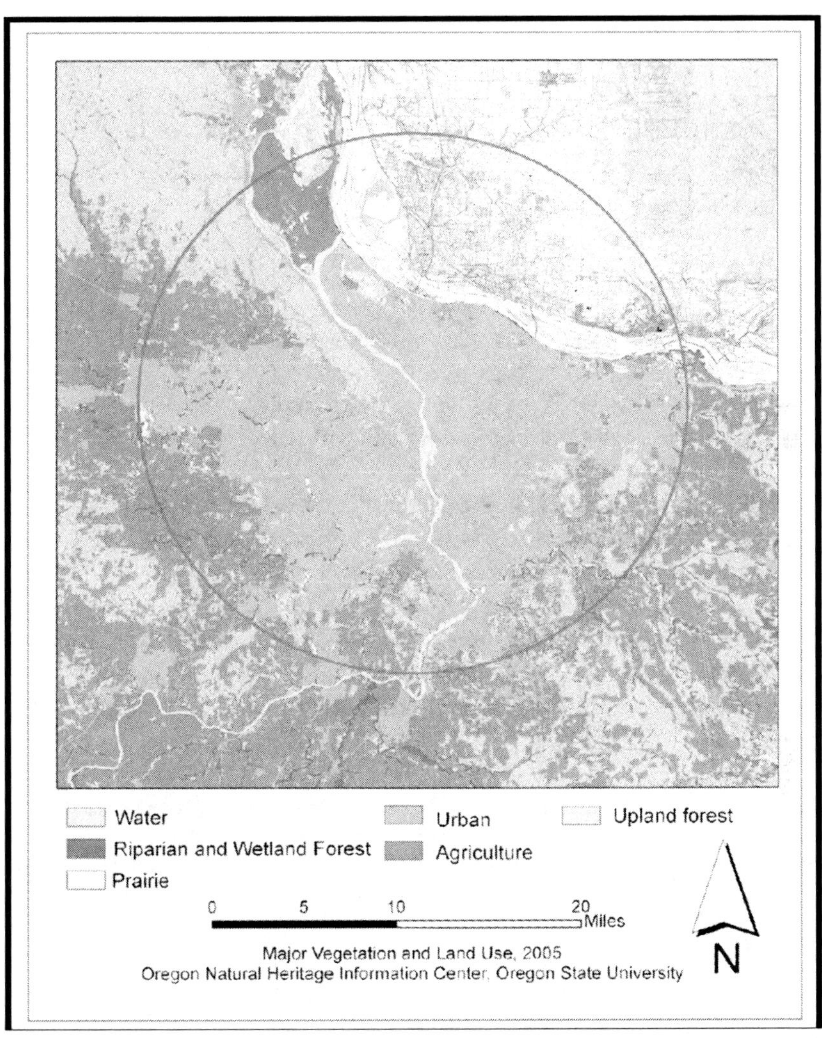

Figure 8.9. Urban land in Portland, Oregon, 2005, when the landscape is primarily urban (concrete and buildings). What do we want the landscape to look like in 2050? Source: Christy, J.A., A. Kimpo, V. Marttala, P.K. Gaddis & N.L. Christy. 2009. Urbanizing flora of Portland, Oregon, 1806-2008. *Native Plant Society of Oregon Occasional Paper* 3: p.34

References Cited for Chapter 8
The Built World

American Society of Landscape Architects (ASLA). (n.d.). Retrieved from http://asla.org

Arratia, Ramon. (2010). Mission Zero: The Power of a Challenging vision. The Endless Possibilities Series. Retrieved from http://www.interfacecutthefluff.com/wp-content/uploads/2010/05/MissionZero.pdf

Aspen Aerogels. (n.d.). Highly Insulating Windows With a U-Value Less than 0.6 W/m2K. Retrieved from http://sites.energetics.com/buildingenvelope/pdfs/aspen.pdf

DesignBoom. (2010). Tiny Houses. Retrieved from http://www.designboom.com/contemporary/tiny_houses.html

Christy, J.A., A. Kimpo, V. Marttala, P.K. Gaddis & N.L. Christy. (2009). Urbanizing flora of Portland, Oregon, 1806-2008. *Native Plant Society of Oregon Occasional Paper* 3: 1-319.

Cyr, Kevin. (n.d.). Small Mobile Homes: Bike Trailers & Shopping Cart Campers. Dornob: Design Ideas Daily. Retrieved from http://dornob.com/small-mobile-homes-bike-trailers-shopping-cart-campers/#ixzz35tjhi6Eq

EcoDistricts. (2013). Retrieved from http://ecodistricts.org/

Edwards, Andres. (2005). *The Sustainability Revolution: Portrait of a Paradigm Shift*. Gabriola Island, BC, Canada: New Society Publishers.

Edwards, Andres. (2010). *Thriving Beyond Sustainability: Pathways to a Resilient Society*. Gabriola Island, BC, Canada: New Society Publishers.

Furuto, Alison. (2012). Tricycle House and Tricycle Garden / People's Architecture Office (PAO) + People's Industrial Design Office (PIDO). Retrieved from http://www.archdaily.com/?p=312651

Giegerich, Andy. (2012). *Portland carbon emissions down 26 percent since 1990. Oregon Sustainable Business*. Retrieved from http://sustainablebusinessoregon.com/articles/2012/04/portland-carbon-emissions-down-26.html

Green Infrastructure Digest. (n.d.). Retrieved from http://hpigreen.com/tag/green-streets/

Greensburg. (2013). Authentic Sustainable Community: Experience the Story: See How We Put the "Green" in Greensburg. Retrieved from http://www.greensburgks.org/

Hawken, Paul, Amory Lovins, and L. Hunter Lovins. (1999). *Natural Capitalism*. New York, NY: Little Brown and Company.

International Living Future Institute (ILFI). (2010). Retrieved from https://ilbi.org/about/faq#how-many-living-building

Lahti, James, and Sarah Torbjörn. (2004). *The Natural Step for Communities: How Cities and Towns can Change to Sustainable Practices*. Gabriola Island, BC, Canada: New Society Publishers.

Leader, Jessica. (2012). Jay Shafer, Tiny House Owner, Gives A Tour Of His 106-Square-Foot Home. Huffington Post. Retrieved from http://www.huffingtonpost.com/2012/12/21/jay-shafer-tiny-house-national-geographic_n_2346023.html

LEED. (2013). Retrieved from http://www.usgbc.org/leed

Living Building Challenge 2.1. (n.d.). Retrieved from http://living-future.org/sites/default/files/LBC/LBC_Documents/LBC%202_1%2012-0501.pdf

National Charrette Institute (NCI). (2011). Retrieved from http://www.charretteinstitute.org/

New Urbanism. (n.d.). Retrieved from http://www.newurbanism.org/newurbanism.html

Novatlantis. (2007). Two Thousand Watt Society: Vision. Retrieved from http://www.novatlantis.ch/en/2000-watt-society.html

One Planet Living. (2013). Ten Guiding Principles. Retrieved from http://www.oneplanetliving.net/what-is-one-planet-living/the-ten-principles/

Register, Richard. (1987). *Ecocity Berkeley: Building Cities for a Healthy Future.* Berkeley, CA: North Atlantic Books.

Roberts, Tristan. (2011). 'Superwindows' To the Rescue? Retrieved from http://www.greenbuildingadvisor.com/blogs/dept/energy-solutions/superwindows-rescue

Rocky Mountain Institute (RMI). (2013). Amory's Private Residence. Retrieved from http://www.rmi.org/rmi/Amory%27s+Private+Residence

Rocky Mountain Institute (RMI). (2013). Reinventing Fire. Retrieved from http://www.rmi.org/reinventingfire

Roseland, Mark. (2005). *Toward Sustainable Communities: Resources for Citizens and Their Governments*. Gabriola Island, BC, Canada: New Society Publishers.

SGBuild. (2013). For the NYC Brownstone. Retrieved from http://www.sgbuild.com/Thermal.html

Steffen, Alex (editor). (2011). *World Changing: A User's Guide for the 21st Century*. New York, NY: Abrams Books.

Sterner, Carl S. (2013). Research. Retrieved from http://www.carlsterner.com/research/2008_waste_and_city_form.shtml

Thompson, Weber. (2008). Eco-Laboratory. Retrieved from http://weberthompson.com/projects/319

Walljasper, Jay. (2007). *The Great Neighbohood Book: A Do It Yourself Guide to Placemaking*. Gabriola Island, BC, Canada: New Society Publishers.

CHAPTER 9
ENERGY

Seventy-two percent of all electricity consumed in the United States is used by homes, offices, and other buildings (Edwards, 2005; USGBC, 2009). These buildings also use 39% of all energy, and produce 38% of all CO_2 emissions (Edwards, 2005; USGBC, 2009). What are some easy things we can change? Hot water heaters use about 25% of the energy in a home (Steffan, 2011). General Electric has a hybrid heater that uses 62% less energy than a traditional water heater (Steffan, 2011). The world leader in solar water heater installations is China, with millions of systems put in place over the past decade (Steffan, 2011). How is water heated in your home?

It might be beneficial to look to nature for answers. How do trees live? If you look at where they thrive, they only grow where there are enough nutrients, and where water is available. The trees send their roots down into the soil only as far as needed. If the environment is wrong for growth, trees don't try to "tough it out." In winter, trees such as the sugar maple drop their leaves, sealing the stem off to prevent water loss. In comparison, how do humans live? We do whatever we want. We move nutrients and water to where we want to grow our food, even when the environment is not fully supportive of growth. And, we use everything in excess. We need to look more closely at our own environment and determine how we can live more harmoniously with nature, following natural patterns and seasons (Benyus, 1997).

It is not a surprise that trees have the ability to save energy. Any plant that is not efficient with energy and nutrients gets "edited out of the gene pool" (Benyus, 1997). Humans are the same. If we do not learn from our mistakes, and begin to use energy and resources wisely, we may also be edited out of the gene pool!

According to Benyus (1997), nature follows two options: it either maximizes or it optimizes, and both serve a purpose in nature. In a field of annual plants, such as the Zumwalt Prairie in northeastern Oregon, nature's main goal is "throughput" – making plants, then seeds, and then quickly releasing the nutrients back into the soil again. The soil is protected from erosion, and the dead organic matter that accumulates each year as the plants die back in winter adds to the richness of the soil. This is also called "maximizing." Humans already understand how to maximize, since that is our standard economic growth model (Figure 9.1). We very quickly turn natural resources into products that are quickly discarded. In many cases our products are made of new materials, unfamiliar to nature. They do not biodegrade and return to the soil. We need to learn how to "optimize." When nature optimizes, the bulk of the plant mass is kept alive, the plants share space, and growth rates slow, as in the Douglas fir forests (Benyus, 1997). Optimizing requires that we change our way of thinking to realize that resources are not limitless. Remember, nature has no landfill for its waste products. Our manufacturing systems have to be rethought so that they also have no waste, and that energy is used wisely.

Figure 9.1. Our current economic model "maximizes" throughput by converting forests into paper that is made into an unread magazine or advertisement, and that is quickly discarded. (Upper left) Source: Bureau of Labor StatisticsUS-Trade.doc; (Upper right) Rolls of paper waiting for to be printed. Source: Bigstock.com. (Lower left) Magazines. Source: iStock.com. (Lower right) Paper waste ready to be recycled. Source: Bigstock.com.

Most of our energy still comes from fossil fuels: oil, coal, and natural gas. In 1956, M. King Hubbert, a geoscientist, predicted that we would reach "peak oil" in the United States between 1965 and 1970. Peak oil means that we have found all the easy oil, and everything from that point on is difficult to obtain. Peak actually occurred between 1970 and 1971. It is time we prepared for a post-petroleum age.

Oil provides 40% of our total energy (transportation, heating, electricity, etc.). We have to import most of our oil from Canada, Mexico, Saudi Arabia, Venezuela, Iraq, Nigeria, Angola, and Algeria (Edwards, 2010). Coal provides more than half of our electricity. We have recently seen the expensive external costs of coal, oil, and nuclear energy in the Upper Big Branch coal mine explosion, the Deepwater Horizon oil spill in the Gulf of Mexico, and in the Fukushima nuclear crisis in Japan. The Japanese power company Tepco continues to paint a rosy picture while radiation levels in its three reactors' surrounding soil and storage tanks are the highest they have ever been (as of September, 2013) since the earthquake and tsunami that caused the meltdown of the reactors. Radiation is leaking into the groundwater and the ocean (Renter, 2013). Four-year-old Albacore tuna have spent considerable time swimming in the radioactive plume from the disaster. Radiation levels measured in fish caught off the coast of Oregon have tripled since 2008 (before the accident) (Delvan and Phillips, 2014; RT: Question More, 2014). Our past energy choices have not been ideal for the health of the planet (Figure 9.2).

Figure 9.2. Our energy options. When we have clean, renewable options, it is curious how some still cling to dirty, destructive options that are clearly bad for the planet and our health, as when people complain, "Not in my backyard" or "Not within my view" for the wind turbines. Source: Bigstock.com.

The ramifications of using fossil fuels have been seen in the increase in carbon dioxide in the atmosphere since the beginning of the industrial revolution (the 1750s). At that time, CO_2 levels were at 280 ppm (parts per million), approximately the same place they had been for the past 420,000 years (Edwards, 2010). Nobel Prize winner Rajendra Pahauri said in 2010, "If there's no action before 2012, that's too late. What we do in the next 2 to 3 years will determine our future!" (Edwards, 2010). We have already begun to see stronger storms and persistent weather patterns like extreme droughts. Solutions for both climate and population stabilization already exist; it is the unfortunate lack of political will that has prevented progress. Even the 2012 (Rio+20) summit in Rio de Janeiro clearly indicated that change was not going to happen at the governmental level. In a Rio+20 summary report, the ISSP (the world's leading professional association of sustainability practitioners) wrote: "Once again, our governments have failed to demonstrate leadership, have lacked courage to make the compromises necessary to ensure a fairer, more stable world" (ISSP, 2014). Who do we count on? Does it have to come down to the individual level?

In the "Do the Math" tour by environmentalist Bill McKibben, he states that there are three numbers we need to remember: 2; 565; and 2,795. Two represents the maximum increase in degrees Celsius for the global temperature; scientists everywhere have agreed on this limit. The number 565 equates to the maximum gigatons of carbon dioxide that can be released and still keep us below 2°C of global temperature increase. The final number, 2,795, represents the gigatons of carbon dioxide currently in the fossil-fuel industry fuel reserves, or 5 times the amount that we can safely burn and not lose control of the weather. At our current rate of fossil fuel consumption, the year 2028 represents the year we will burn through the 565 gigaton allotment (Fleischer, 2012). We, as individuals, need to find every possible way to decouple from our dependence on fossil fuels. We can't count on the government or corporations to do this for us.

According to Hawken et al., (1999):

- 50% of the threat to the climate comes from burning fossil fuels;
- 25% of carbon threat is due to CO_2 released from soil erosion, logging, poor grazing, farming, or ranching practices; and
- The remaining 25% vanishes if CFCs are replaced with substitutes already required by global agreement but not enforced.

So it seems that all that is required is to embrace energy efficiency in all its forms: use better agricultural practices; enforce laws already passed by global agreement but not enforced; and change to renewable energy forms. To address these concerns, the Carbon Mitigation Initiative (CMI) aims to reduce carbon emissions by 1 billion tons (907.2 million t) per year until 2054 (Edwards, 2010). Their systematic approach addresses the complex problem through 15 strategies. You can attempt to address the problem of emissions yourself, using these 15 strategies in a simulation game on the CMI website (CMI, 2011).

We first need to reduce the amount of energy we consume: conservation. That requires no new technology, no increased expenditure for infrastructure, we just need to pay attention to patterns and change them. Every day, ask "How can we change or reduce our energy consumption?" and "How can I change or reduce my energy consumption?"

The amount we spend as a nation to light up the night sky must be enormous. There was a time when we could see the stars, but artificial light now makes even the brightest of stars dim in our night sky. Images from space (Figure 9.3) document this extreme waste of energy. Imagine 50% more lights added to this image as our population grows. The expected growth of light pollution in the United States as well as around the world, has been well documented by P. Cinzano (2002) at the Istituto di Scienza e Tecnologia dell'Inquinamento Luminoso of Thiene, Italy. His dramatic series of images from 1950, 1970, 1997, and projecting to 2025 show the excessive use of night sky illumination in our country (Cinzano et al., 2001). The 4-map image comparisons can be seen online at http://www.sierranaturenotes.com/naturenotes/NALightPollution.htm

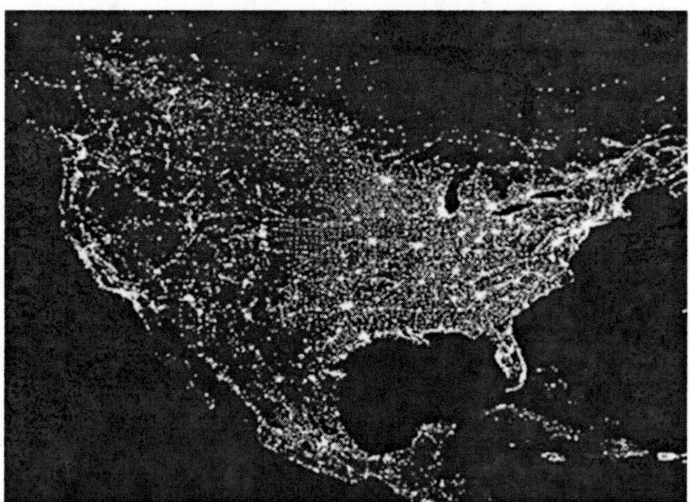

Figure 9.3. Night sky illumination of the United States in 2008. Source: NASA/GSFC/Craig Mayhew and Robert Simmon.

My son went on an archeological dig in Jordan after he graduated from high school. I was certain he would be amazed at the beauty of the night sky in the middle of the desert in Jordan. But an engineer had sold sodium vapor highway lights to the country, even though they were not necessary, and their night sky was no different there than in our urban environment. The pattern of lights from each city in the world has its own unique signature from space, due to geography, culture, and the technology available, as documented by the astronauts at the International Space Station (Johnson, 2014). How much energy could be saved if only 50% of the lights were removed?

Could we do with less light at night? While I was in school in Cambridge, Massachusetts, a truck hit a relay station of some kind, and the streetlights were knocked out. The lights within all the buildings remained intact. What happened overall? Cars drove more slowly, which we need anyway to improve pedestrian safety in our neighborhoods. People carried flashlights, walked closer, and held onto each other, which was very sweet. The lights shining out from the stores and homes seemed animated and beautiful. The whole experience was magical. Can we rethink where we actually need night light and reduce this to a more reasonable level?

Every year we have “Earth Hour,” when we turn off the lights for one hour. In 2007, 2.2 million people participated around the world. In 2008, 50 million people turned off their lights in 35 cities officially registered for the Earth Hour event and 400 unofficial cities. In 2009, hundreds of millions of people in 4,000 cities and 88 countries participated (Edwards, 2005). If we can do it for an hour, can we consider doing it longer? Why do we have to have so many lights on in our cities (Figure 9.4)? Take back the night sky!

Night illumination is actually a form of pollution. Our exposure to bright nocturnal light causes a decrease in our ability to produce melatonin, a hormone required to regulate our sleeping and waking cycles. Reduced melatonin production has been linked to reproductive cancers (prostate cancer in men and breast cancer in women) (Light Pollution, 2013). Should we not consider turning off the lights more than just one hour to save energy *and* our health?

Figure 9.4. (Left) New York City night illumination. Source: iStock.com. (Right) Light Pollution, Tenerife, Canary Islands. Source: Jose Angel/Creative Commons via flickr.

The question is whether we can produce enough energy using just green sources. MUST we rely on coal, oil and nuclear? There are cleaner, safer options available and ready right now. According to a report by the Union of Concerned Scientists (UCS) in 2009, we can meet our energy demands for the next 20 years without coal-fired power plants or nuclear reactors. We can increase efficiency, and increase our use of wind and solar sources. This would reduce emissions by about 85% by 2030 and would be a great step in the right direction. The full report is available online at the UCS website (UCS, 2009).

Energy Upgrade in California is a world model for an energy-efficiency program. Its goal is for the state to reduce per capita energy use 40% below the national average, while remaining the eighth largest economy in the world (Edwards, 2010). Its mission is to develop better energy efficiency, "streamline energy use in agriculture, manufacturing, water systems, and processing functions"

(Edwards, 2010), and to help everyone implement better efficiency measures across the state.

Corporations are also finding that conservation pays. By using energy-efficiency measures, Dow Chemical saved 4 times more than expected, an amount of 7 billion dollars, which also avoided the emission of 70 million tons of carbon dioxide (63.5 million t) (Davis, 2008). Over a longer period of time (1994-2010) Dow prevented the emission of 104.7 million tons (95 million t) of CO_2 worth 9.4 billion dollars – enough energy to power every residential home in California for a year and a half. The employees at Dow submit ideas for efficiency to the corporation, in the form of a competition, and have submitted over 60 ideas that, if implemented, would eliminate the need for 8 trillion BTUs of energy and cut CO_2 emissions by 440,925 tons (400,000 t) (Bardelline, 2011).

People *assume* that carbon abatement will be too costly. More realistic assumptions include the fact that technological breakthroughs have already happened, and we could reduce our energy bills by at least 300 billion dollars a year using the technologies we already have to provide better, cleaner services (Hawken et al., 1999). Why are we stalling when we can protect the Earth, and not at a cost, but at a profit?

For renewable energies, wind energy capacity (the sustained output of facilities) has increased tenfold between 2000 and 2008 in the United States, surpassing Germany in wind energy, but not solar. Renewable energy accounts for only 2.5% of the energy we generate. Obviously, conservation will need to be a big part of reducing the other 97.5% of our energy usage until the alternatives are in place at a larger scale. Conservation is essential, yet it is still often overlooked (Edwards, 2010).

A windfarm built off the coast of Great Britain was expected to have 175 of 341 planned turbines operational before the London Olympics in 2012. An array of 175 turbines did finally open in July, 2013, and now powers half a million homes. Using these figures, one turbine would power 2,800 homes (Pinsent Masons, 2013; Steffan, 2011). How many homes do you have in your neighborhood? Maybe you only need 1 large turbine or 2 to 3 small turbines to power your entire neighborhood. Where might these turbines be placed? Battelle Pacific Northwest National Laboratory in western Washington says it could very quickly supply 20% of the total electricity needs in the U.S., and wind could supply one third of our energy by 2020! The total wind energy that is possible could generate 15 times more energy than what the world uses currently (Steffan, 2011). Many wind farms are planned, under construction, or are already built in the Columbia River Gorge.

Is there enough wind in other parts of the state for personal or neighborhood turbines? The National Renewable Energy Lab (NREL) publishes a map of average wind speed at 80 meters (Figure 9.5). The new turbines can operate at a speed of 4 miles per hour. Even at that low speed, most of Oregon definitely has enough wind to merit consideration of more turbines. A micro-wind turbine (25 centimeter diameter rotor, or about a foot) was in development for use in both rural and urban environments (Levesque, 2007). It could generate power with as little as 4.5 miles per hour (two meters per second). It is difficult to find an updated report on this style of turbine, but Clean Technica (2013) has evaluated others. Wind turbines are becoming more efficient, quieter, and can operate at lower and lower wind speeds. Urban environments tend to have low wind speeds because of the trees and tall buildings. At an individual home scale, it may be possible to design and develop other versions that work in urban settings with recycled materials.

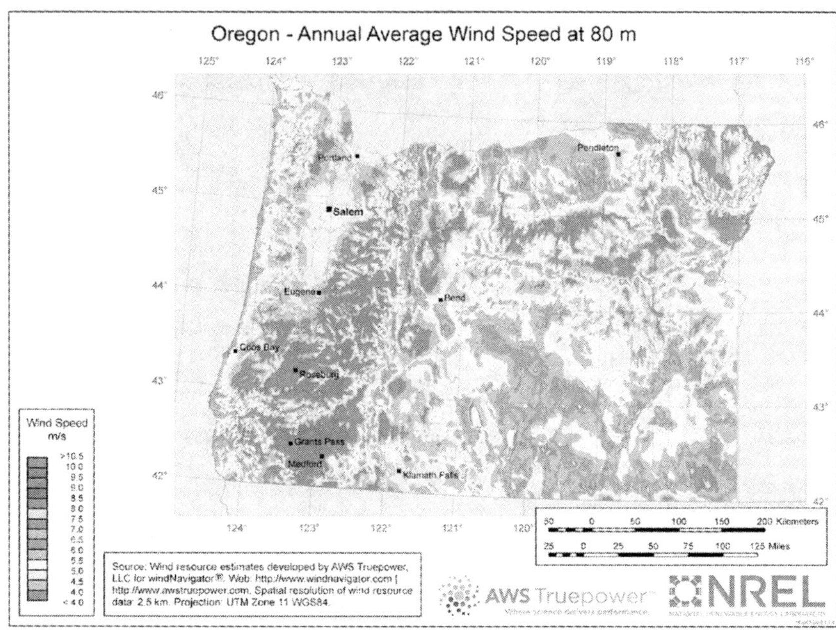

Figure 9.5. Average wind speeds at 80 meters. Source: National Renewable Energy Laboratory. www.windpoweringamerica.gov/wind_resource_maps.asp?stateab

Michael Davis (2006-2012) published online directions for building a wind turbine, including detailed photographs at every step. In 2007, it was built for less than $150. Davis claims it to have a possible worth of $1,000. Davis' directions are very informative and quite humorous:

> I never got a chance to properly test the unit before heading to Arizona. One windy day though, I did take the head outside and hold it high up in the air above my head into the wind just to see if the blades would spin it as well as I had hoped. Spin it they did. In a matter of a few seconds it spun

> up to a truly scary speed (no load on the generator), and I found myself holding onto a giant, spinning, whirligig of death, with no idea how to put it down without getting myself chopped to bits. Fortunately, I did eventually manage to turn it out of the wind and slow it down to a non-lethal speed. I won't make that mistake again.
>
> Michael Davis (2006-2012)

Portland, Oregon, is one of the cities that has set targets and strategies for renewable energy. A Community Energy Planning Tool (CEPT) is available to help communities compare various design options in order to be more efficient and sustainable (CEPT, 2008). Tulsa, Oklahoma, features a system of off-the-shelf turbines that recapture waste heat and power which heats and cools the downtown area. If this one innovation were adopted across the country, CO_2 emissions could be reduced 23% nationwide (Hawken et al., 1999).

When considering energy efficiency, again, biomimicry enters the picture. Biomimicry scientists have studied termite mounds in Zimbabwe that maintain an even temperature year-round. By emulating the mounds' structure in buildings, fresh air is drawn in at the ground level, where fans push the air to the building core. The fresh air replaces the stale, hot air which rises and exits through passageways in the ceilings of each level. Buildings designed in this fashion use 10% of the energy used by a conventional building, saving, as in this example, 3.5 million dollars per year on the cost of air conditioning alone (Edwards, 2005). Not only is this a more efficient way to cool our buildings, but heating in this manner would also be more efficient.

Renewable technologies are the fastest growing technologies, with wind increasing 26% each year, and solar growth increasing 23-43% (Hawken et al., 1999). What we each need to consider is how to bring efficient energy generation to the community level.

It is at the small scale, at the local level, that the area with the greatest potential for many unique innovations is found. We have spent all of our resources in the past building bigger and bigger systems. To reverse this trend it will take decades to decentralize and diversify our power sources. We are finding that decentralized power sources are ten times more valuable that anticipated. That makes solar cells cost effective even now (Hawken et al., 1999). In Indiana, it was discovered that it was much more effective to produce energy at the scale of one small energy plant for every three city blocks to eliminate the power losses normally seen between power generation source and power usage (McDonough and Braungart, 2002).

New alternative sources for energy may also be on the horizon. Algae, not a plant but a protist, is being grown and converted into biodiesel, a biofuel. The fuel yield from algae is dramatically higher than from any other source (Figure 9.6) (Goffman, 2009). Human urine has also been linked to these energy systems as a liquid fertilizer (Grunbaum, 2010). Researchers are beginning to link systems, just as nature does, by combining waste systems with energy systems.

More than one third of the energy used in a building can be saved at no extra charge. Passive systems take advantage of solar gain, and buffer against the wind or against undesired heat. Energy savings can be accomplished using thermal mass, shading, and passive-solar gains, etc. Access to passive light can be increased through light tubes, light pipes, light-colored surfaces, and curved light shelves to bounce additional light into the interior spaces of homes and buildings (Hawken et al., 1999).

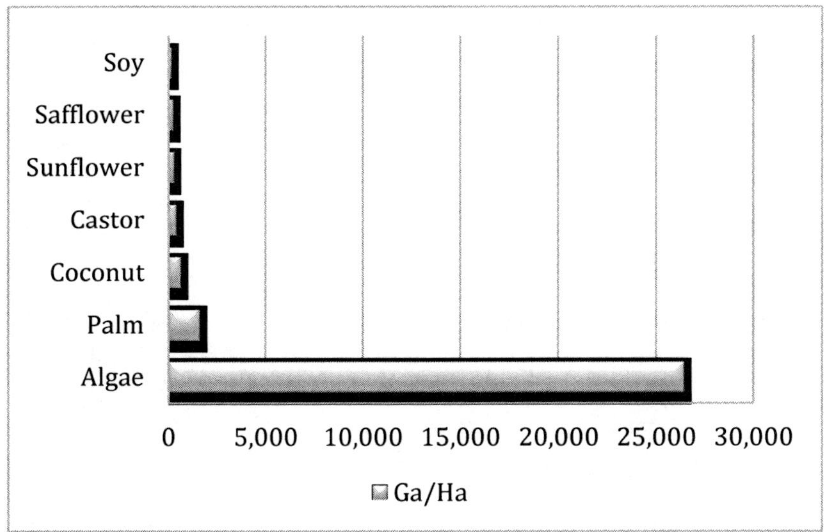

Figure 9.6. Yield of various plant oils and algae (a protist) in gallons per hectare. Source: Goffman, 2009.

Simply unplugging appliances when they are not in use can increase energy efficiency (Steffan, 2011). Many appliances use as much energy when they are off as they do during operation; this is called vampire power or standby power. A good summary of the appliances that are involved in this energy drain is available at Wikipedia (online at http://en.wikipedia.org/wiki/Standby_power). Approximately 20% of our monthly power bill goes to vampire power. Notice that many electronics no longer have an "off" button anymore. It has been replaced by "standby" instead. Some TV's and stereo systems still use 70-80% of full power when turned off (Steffan, 2011). Plug these appliances into a power strip that can easily be turned off, disconnecting the appliances and the energy drain.

The U.S. lags far behind two European countries when it comes to diverting waste from landfills. Waste destined for a landfill amounts to 69% of all waste in America. In Denmark it is only 3%;

and an astonishing 0% in Germany—*nothing* goes to a landfill in Germany! We send only 7% of our trash to waste-to-energy programs, whereas Denmark sends 54%, and Germany 38% (Giegerich, 2013). Two problems are solved with this single practice: less land needed for landfills, and energy is produced from what would be discarded.

Neighborhood Natural Energy is a non-profit organization in Portland, Oregon, that aims to create small-district energy projects all over the city, varying in size from a few homes to several hundred homes and multiple businesses (Steffan, 2011). The Portland Planning and Sustainability Bureau describes in detail this new form of energy structure (District Energy, 2013). There is also a report explaining how you can increase solar energy in your community (Solar America Community, 2012).

Linked with energy conservation, there are other ways to reduce CO_2 emissions. Steffan (2011) reports that coal fires (underground fires that burn in the coal seams and mines for years if not decades) contribute an enormous amount of CO_2 emissions. If the coal fires in China alone could be extinguished, it would have the same result as removing all the cars in the United States from the highway (Steffan, 2011). If we find a way to extinguish these fires, we buy ourselves a great deal of time in the energy field.

In other parts of the world, embracing green energy has a different meaning. Villagers in places that have no energy, or only energy from burning firewood, find their lives tremendously uplifted with the addition of solar panels to their village. Illiterate women and unemployed youths are selected and trained by an organization called "Barefoot College" with the skills needed to install and service solar panels in their respective communities. The trainees return to their villages as new community leaders. All of this is accomplished without elaborate textbooks (Steffan, 2011).

A typical child in the Global South does his or her reading by the light of a kerosene lamp. In a single evening, the child breathes in enough fumes to equal the smoking of two packs of cigarettes (Steffan, 2011). For heating and cooking, gathering firewood is required; this takes a lot of time in rural areas, and is very expensive in urban areas. The smoke pollutes the air and contributes to global warming (Steffan, 2011). Now, a single white LED light (called d.light) run from a battery powered by solar panels, can give enough light to read by. Soon these lights will be attached to every kind of building material in the homes of billions of people around the world (Steffan, 2011).

Power from the sun can be used and stored in many forms. A solar heating system called SolSource 3-in-1 was designed for villages in the Himalayas. Shaped like an upside-down umbrella, the SolSource 3-in-1 is made from local materials: fabric is stretched over a bamboo frame and lined with a reflective material that captures sunlight, concentrating the light's energy in the center of the apparatus. At this central position, you have several choices:

- You can place a pot for cooking;
- You can use a device that generates electricity and stores energy for seven hours;
- You can connect a heating module to heat the home;
- You can attach a disinfector to treat drinking water; and
- You can activate a thermal battery to allow cooking after dark, heating water in 5 to 7 minutes (Steffan, 2011).

Available only in the Global South, the SolSource 3-in-1 would also be an excellent emergency power system in any home, or a permanent system in any off-the-grid home.

How should we address energy in our communities? Maybe a first step is to consider having a contest. Such an official contest is underway for Portland's office buildings. Why not have one between residences? Or neighborhoods? Or within an apartment building? Americans may find it easier to strive to be "better than" another group, rather than just striving to be better as isolated individuals. Start a competition yourself!

As a final question, why does it matter that we change the form of energy we use? In his inspiring article, *Al Gore explains why he's optimistic about stopping global warming*, Klein (2013) puts into perspective just how much energy we are wasting in the form of carbon dioxide being released into the atmosphere. He states that every day the energy that is trapped due to manmade global warming is equal to 400,000 Hiroshima bombs exploding every 24 hours, and although the planet is pretty big, that is a lot of energy! There is no doubt that there are better ways to use that energy than to throw it into the atmosphere. How we decide to change, and what form that will take has the potential to be pretty amazing in itself. You are the ones that are going to make this change. You just don't know it yet!

References Cited for Chapter 9
Energy

Bardelline, Jonathan. (2011). *Dow Puts $100M Towards Employee Energy Efficiency Ideas.* Retrieved from http://www.greenbiz.com/news/2011/03/02/dow-puts-100m-towards-employee-energy-efficiency-ideas

Benyus, Janine. (1997). *Biomimicry: Innovation Inspired by Nature.* New York, NY: HarperCollins Publishers Inc.

Carbon Mitigation Initiative (CMI). (2011). Wedges Game, Retrieved from http://cmi.princeton.edu/wedges/game.php

Cinzano, P., F. Falchi, and C.D. Elvidge. (2001). Artificial night sky brightness due to light pollution in North America: A preliminary picture of the growth from 1950 to 2025. Retrieved from http://www.sierranaturenotes.com/naturenotes/NALightPollution.htm

Cinzano, Pierantonio. (2002). *The growth of the artificial night sky brightness in North America in the period 1947-2000: a preliminary picture, in Light Pollution: a Global View*. ed. H. Schwarz. p. 39-48 (ISBN 1-4020- 1174-1). Dordrecht: Kluwer.

Clean Technica. (2013). *Reshuffling Our Top 5 Micro Turbines.* Retrieved from http://cleantechnica.com/2013/04/30/the-5-best-micro-wind-turbines-2013/

Community Energy Planning Tool (CEPT). (2008). Oregon Department of Energy. Retrieved from http://www.oregon.gov/ENERGY/GBL WRM/docs/CommunityEnergyPlanningTool.pdf

Davis, Michael. (2006-2012). *How I home-built an electricity producing Wind turbine*, Retrieved September 17, 2013 from http://www.mdpub.com/Wind_Turbine/

Davis, Valerie. (2008). Dow Chemical Ties Energy-Efficiency to Cash Savings, Retrieved from http://www.environmentalleader.com /2008/12/10/dow-chemical-ties-energy-efficiency-to-cash-savings/

Delvan, Neville and Jason Phillips. (2014). Study finds only trace levels of radiation from Fukushima in albacore. News and Research Communications. Oregon State University. Retrieved from http://oregonstate.edu/ua/ncs/archives/2014/apr/study-finds-only-trace-levels-radiation-fukushima-albacore

District Energy. (2013). Neighborhood Scale Development: What is District Energy? Retrieved from http://www.portlandoregon.gov/bps/article/349437

Edwards, Andres. (2005). *The Sustainability Revolution: Portrait of a Paradigm Shift.* Gabriola Island, BC, Canada: New Society Publishers.

Edwards, Andres. (2010). *Thriving Beyond Sustainability: Pathways to a Resilient Society.* Gabriola Island, BC, Canada: New Society Publishers.

Fleischer, Matthew. (2012). Bill McKibben's 'Do The Math' tour points to 2028 as the year of catastrophic climate change. Huff Post Green. Retrieved from http://www.huffingtonpost.com/2012/11/13/billmckibben -do-the-math-2028_n_2123406.html?view=print&comm_ref=false

Giegerich, Andy. (2013). Covanta amps waste-to-energy efforts, re-ups Oregon contract. *Oregon Sustainable Business*. Retrieved from http://www.sustainablebusinessoregon.com/articles/2013/09/covanta-amps-waste-to-energy-efforts.html

Goffman, Ethan, 2009. Biofuels: What Place in Our Energy Future, Algae: Power from Pond Scum. Retrieved from http://www.csa.com/discoveryguides/biofuel/review7.php

Grunbaum, Mara, 2010, July 23. Gee Whiz: Human Urine Is Shown to Be an Effective Agricultural Fertilizer. *Scientific American*. Retrieved from http://www.scientificamerican.com/article/human-urine-is-an-effective-fertilizer/

ISPP. (2014). Disappointment Hangs Over Conference Outcome As Governments Exhibit Lack of Resolve (July Newsletter Article). Retrieved from https://www.sustainabilityprofessionals.org/summary-reporting-about-rio20-2012-conference-issp

Johnson, Terrell. (2014). 21 Awe-Inspiring Photos of Cities at Night From Space, Taken by Astronauts on the International Space Station. Retrieved from http://www.weather.com/news/science/space/21-awe-inspiring-satellite-images-cities-night-around-world-20140717

Klein, Ezra. (2013). *Al Gore explains why he's optimistic about stopping global warming*. Retrieved from http://www.washingtonpost.com/blogs/wonkblog/wp/2013/08/21/al-gore-explains-why-hes-optimistic-about-stopping-global-warming/

Levesque, Tylene. (2007). *Micro Wind Turbines: Small Size, Big Impact*. Retrieved from http://inhabitat.com/micro-wind-turbines-small-size-big-impact/

Light Pollution and Health. (2013). Retrieved from http://physics.fau.edu/observatory/lightpol-BrCa.html

McDonough, William and Michael Braungart. (2002). *Cradle to Cradle: Remaking the Way We Make Things*. New York, NY: North Point Press.

Pinsent Masons (LLP). (2013). World's largest offshore windfarm opens off the coast of Kent. Retrieved from http://www.out-law.com/articles/2013/july/worlds-largest-offshore-windfarm-opens-off-the-coast-of-kent/

Portland Kilowatt Countdown. (2013). Retrieved from http://kilowattcrackdown.betterbricks.com/portland
Renter, Elizabeth. (2013), September. *Radiation Levels Hit 'New High' at Fukushima*. Natural Society News, Retrieved from http://www.nationofchange.org/radiation-levels-hit-new-high-fukushima-1378647935

RT: Question More. (2014). Radiation level in tuna off Oregon coast triples after Fukushima disaster. Retrieved from http://rt.com/usa/155692-oregon-tuna-radiation-tripled-fukushima/

Solar Cooker Review. (2009). Retrieved from http://www.solarcooking.org/newsletters/scrjul09.htm

Solar Powering Your Community, a Guide for Local Governments, U.S. Department of Energy. (2012). Retrieved from http://solaramericacommunities.energy.gov/resources/guide_for_local_governments/

Steffen, Alex (editor). (2011). *World Changing: A User's Guide for the 21st Century*. New York, NY: Abrams Books.

Union of Concerned Scientists. (2009). Climate 2030: A National Blueprint for a Clean Energy Economy Retrieved from http://www.ucsusa.org/publications/ask/2011/coal-nuclear.html

U.S. Green Building Concil (USGBC), South Carolina Chapter. (2009). What is Green Building? Retrieved from http://www.usgbcsc.org/site/?page_id=140

CHAPTER 10
PEOPLE, CULTURE AND COMMUNITY

In becoming a sustainable nation, a sustainable world, we are the problem. People are the problem. The conservation issues, the technical problems, they can all be solved—but not before we deal with the "people problem." That has to come first. If we are going to understand and fully support our communities, if we are going to live a truly sustainable life, in harmony with everything in the world, then we have to change the way we do *everything*. We need to respect the needs of everyone, to support all people within our communities, and within all cultures. Any other path will not be sustainable in the long run.

This understanding of having to start with the individual to attain anything on a large scale is summed up well in this 6th century BCE poem by the Chinese poet Lao-Tse, entitled "Peace."

> If there is to be peace in the world,
> There must be peace in the nations.
> If there is to be peace in the nations,
> There must be peace in the cities.
> If there is to be peace in the cities,
> There must be peace between neighbors.
> If there is to be peace between neighbors,
> There must be peace in the home.

> If there is to be peace in the home,
> There must be peace in the heart.
>
> World Prayers, n.d.

The biggest sustainability challenge facing the developing world is that of helping people escape poverty. In the developed world, the issue is to make prosperity sustainable (Steffan, 2011).

We have to embrace culture. What is culture? Often it is invisible to the naked eye. It is hard to describe. Frequently, it is just "the way we do things around here" (Doppelt, 2003). It goes beyond race and gender. It has to do with our perspective (Figure 10.1). We are each many different people, depending upon the situation, or

Figure 10.1. Culture goes beyond race and gender to *perspective* … in the set of images on the left a woman is about to fall in the water, or has she already done it? In the set on the right, are the athletes divers or gymnasts? Source: Bigstock.com.

upon different times in our lives (Steffan, 2011). If we embrace greater compassion of culture and diversity, greater creativity will follow.

We also have to embrace gender issues. When third-world women leave their farms and move into cities, they become free to make more of their own choices; they have access to better jobs and education, and to healthcare. Under these conditions they almost always choose to have fewer children. It becomes clear that the key to living sustainably around the world is tied to women's rights (Steffan, 2011). In Mumbai, India, women came together to transform their neighborhood. As a result, instead of homes being built of mud, they are now concrete; pathways are tile to prevent erosion; the neighborhood even pooled their resources so that they would have their own bank. "Instead of agonizing in the face of hardship, they organized" (Steffen, 2011). What can we learn from them? Don't agonize! Organize!

Gender equality is a worldwide issue. You have to wonder why the United States does not even feature in the top 20 countries ranked by equality or fairness to women (Figure 10.2). Canada is number 21 and the United States number 22 in the data presented by Cann (2012), the "Ranking of Countries and Gender Gap."

Just as we need to strive for gender equality, we also need to embrace our senior citizens. Very few places are really ready for the wave of senior citizens soon to populate our communities. By 2050 there will be 19 million people in the United States who will be over 85 years of age (Steffan, 2011). How will we care for our elders? What place should they have in our communities? Companies will benefit when they grasp the qualities that senior citizens have to offer. A research study performed by the SHTM Survey Program lists the many benefits of hiring older workers (Figure 10.3).

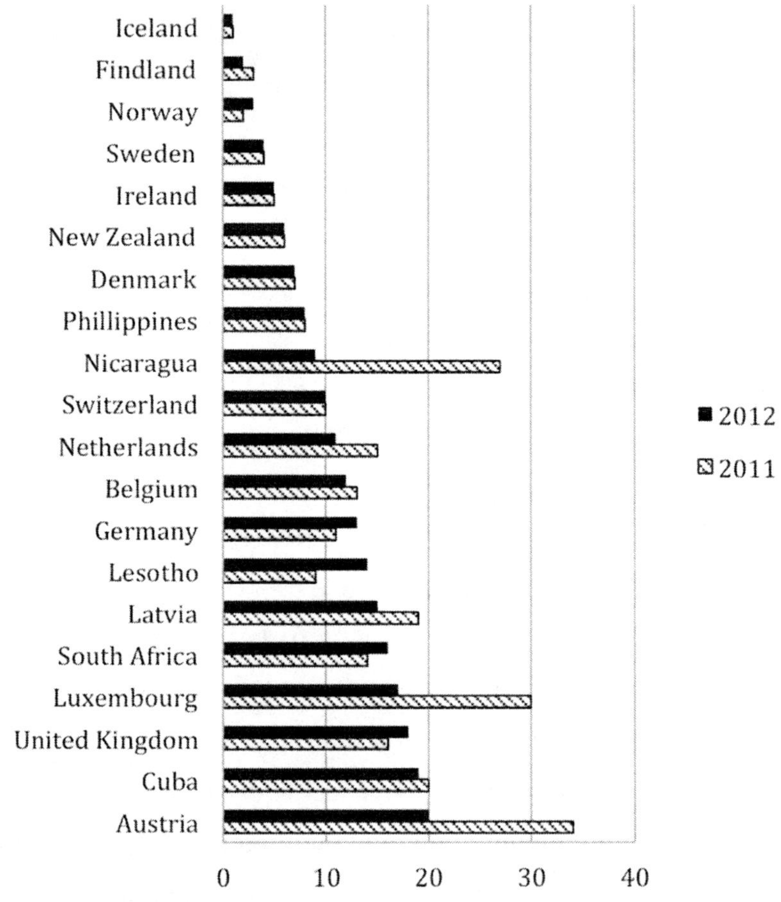

Figure 10.2. The countries with data extending beyond the basic range of the data have made significant progress in women's rights. If the gray bar is shorter than the black one, women are losing ground. Where is the United States? Not in the top twenty! Source: Cann (2012).

Benefits of hiring older workers (SHRM, 2003)
☐ Older workers are more willing to work different schedules
☐ Older workers serve as mentors for workers with less experience
☐ Older workers have invaluable experience
☐ Older workers are more reliable
☐ Older workers add diversity of thought/approach to team projects
☐ Older workers are more loyal
☐ Older workers take work more seriously
☐ They have established networks of contacts and clients
☐ Older workers have higher retention rates
☐ Older workers have more knowledge and skills
☐ Older workers are readily available
☐ Older workers are more productive

Figure 10.3. Benefits that senior citizens have to offer in the work place of the future Source: SHRM (2003).

As we embrace our seniors and design our communities to be inclusive of those in older age groups, so must we also engage our youth. Median ages of populations in various countries are listed in Figure 10.4. There are 1.2 billion young people in the world between the ages of 15 and 24, and they can't find work. How can people have a meaningful life if they cannot even support themselves? How can we find ways to provide for other cultures without destroying their uniqueness.

Japan	42
Italy	42
Finland	40
United States	35
Worldwide	28
Least developed countries	19

Figure 10.4. Median age of the population in a sampling of various countries. Source: Steffan (2011).

And what about our American culture? We should no longer want to be homogenous, with a fast food restaurant on every corner. We should want to be different, as an expression of who we are and where we live. The Southwest should look different from upstate New York. How we solve problems or develop our sense of community in Portland, Oregon, should not be the same as how it is done in northern Florida. However, that being said, we *can* learn from each other and then find our own form of expression. So where can we look for inspiration? In a very unlikely place: Curitiba, Brazil. Although this chapter is about people and culture, those are at the heart of all the changes to *everything* in Curitiba.

Curitiba is approximately the same size as Houston, Texas, both in area and in population. Like many cities in undeveloped countries, it has scant resources and explosive population growth. The percentage of its population under 18 years of age is 42%, and another million residents are expected by 2020 (Hawken et al., 1999). While it is well known that cities in the Global South attract poverty, unemployment, squalor, disease, illiteracy, inequity, congestion, pollution, corruption, and despair, somehow Curitiba has achieved just the opposite! In 30 years they developed "better levels of education, health, human welfare, public safety, demographic participation, political integrity, environmental protection and more community spirit than its neighbors, and certainly more than most U.S. cities" (Hawken et al., 1999). All of this was accomplished without generating megaprojects, "just hundreds of multipurpose, cheap, fast, simple, homegrown, people-centered initiatives using common sense and local skills" (Hawken et al., 1999). This was possible because Curitiba treated all of its citizens, especially its children, as something to be treasured.

As city governments can get bogged down in legalities, Curitiba didn't always wait for the approval of all departments. For example, one Friday evening as the businesses closed for the weekend, workers moved in and started jackhammering the pavement. Working around the clock, they laid cobblestones, installed streetlights and kiosks, and planted tens of thousands of flowers. Just 48 hours later, Brazil's first pedestrian mall (one of the first in the world) was ready for business (Hawken et al., 1999). Monday morning, when it was time to open, businesses complained! Cars tried to retake the streets, but the city rolled out huge rolls of paper and had children paint watercolors in the street to prevent the take-over by cars. In remembrance, there is a "paint-in" every Saturday morning (Hawken et al., 1999).

Who was the visionary that transformed Curitiba? The mayor, Jaime Lerner, was chosen because of his lack of political talent. He was a 33-year-old architect, engineer, urban planner, and humanist. He was cheery, informal, energetic, and intensely practical. As mayor he turned out to be charismatic, compassionate, a visionary and the most popular mayor in Brazilian history (Hawken et al., 1999). Curitiba found that innovations were led by women and architects. In other words, by problem solvers rather than nay-saying politicians! Risks were taken, and mistakes were expected.

Curitiba was designed to place the increased density near the existing transportation boulevards. The architects met the needs of the poorest citizens by favoring mass-transit (Figure 10.5) over private cars. The city has the lowest rate of car drivership and the cleanest urban air. Rush hour express buses leave and arrive once a minute! They didn't spend money they did not have. To give cultural meaning to neighborhoods, they created ceremonial gates and centers for each culture to honor the local ethnicities. Lerner revitalized the arts and culture, and protected historic buildings. How can we

Figure 10.5. Jaime Lerner, mayor, architect, and urban planner of Curitiba, Brazil, was the designer of the bus stop "tubes" which hasten loading and unloading of buses. Source: iStock.com.

celebrate culture in our own city? Where should ethnic centers be located? Vandalism in Curitiba is unknown due to their civic pride (Hawken et al., 1999). What will give us civic pride?

Areas of the city that were prone to flooding were transformed into parks and lakes where excess water could be easily absorbed. They planted hundreds of thousands of trees (Hawken et al., 1999). To maintain grassy areas, noisy, smelly, oil-consuming lawn mowers were replaced by a flock of 30 sheep that move around as needed. The wool and meat from the sheep provide income for social programs (Hawken et al., 1999). Goats are popular in Portland. Can we increase the number to maintain public areas and perhaps develop a local brand of goat cheese?

We should be so lucky as to have the recycling rate of Curitiba (Figure 10.6), where 79% of households recycle 3 times a week. The materials that are collected are sold to recover 50% of the operating expenses. How often are our recyclables collected? How are the resources used? Does our local economy benefit from their sale?

Figure 10.6. Color-coded recycling bins in Curitiba. Source: Juliana Swenson.

Many are drawn to Curitiba from all parts of Brazil, and squatter villages are created. The paths between homes are so small that sanitation trucks cannot enter. In order to keep the area clean, small trucks pull up to the hundreds of squatter sectors and drivers ring a bell. Thousands of citizens bring bags of garbage to the truck to swap for food: 60 kilograms of trash will earn 60 food tickets (enough food for a month), or bus tokens, schools supplies, or toys for Christmas. According to Hawken et al., (1999) the coupons say: "You are responsible for this program. Keep on cooperating and we will get a cleaner Curitiba. You are an example to Brazil and even to the rest of the world." The food exchanges give rice, beans, potatoes, onions, oranges, garlic, eggs, bananas, carrots, and honey—the seasonal surplus produce from local farmers. This helps to keep the farmers on their farms. Hard to reach litter is cleared out, and 135 neighborhood associations hire the unemployed or retired people in need (Hawken et al., 1999).

In education, Curitiba has achieved Brazil's highest literacy rates (over 94% in 1996), and lowest first-grade failure rates. Environmental education is taught across all courses and curriculum (Hawken et al., 1999). In comparison, what are the literacy rates in the United States? Nationally, we are at 85.5% and Oregon stands at 90%, both lower than that of Curitiba.

Children are given work. When vandalism did occur, such as when gangs tore up the flowerbeds at the new Botanical Garden, it was interpreted as a cry for help, and the gang members were given jobs as assistant gardeners. Older children are given apprenticeships in forestry, environmental restoration, water pollution control, and public health (Hawken et al., 1999).

A survey taken in the 1990s documented that 99% of Curitibans would not want to live anywhere else. In nearby San Paulo, 70% of citizens thought life would be better in Curitiba. And 60% of New Yorkers would leave the glitter of their city to move to Curitiba (Hawken et al., 1999). Instead of moving, let's improve our own neighborhoods!

The average income in 1996 for Curitiba was $7,827 and the average American income was $28,988. With incomes 73% lower than ours, Curitibans have created, says Bill McKibben, one of the world's great cities (Moore, 2007). Perhaps we need to look more closely at how our money is spent. Can we get more value from the funds we already have available in our cities and communities? Can we get more value from the funds we already have available in our own personal lives?

If the human factor is the most difficult to change, how do we tackle this problem? Luckily we are not the only ones looking for answers. Many cities have sustainability offices. In Portland, it is the

Portland Bureau of Planning and Sustainability. Many companies within cities are stepping up to the plate and are looking for sustainable options in their line of work. Both of these approaches are from the "top down." They commence in large organizations and filter down to the individual. It is essential that inspiration and motivation also begin at the individual level and work from the bottom up as well.

With the world's population increasing at the rate of 211,000 more people per day, new individuals must be born into a sustainable way of life, and not follow in our old footsteps (Figure 10.7). In general, the undeveloped world wants to live as the developed world does, but it is not true that they need to copy us. They don't need to play catch-up (Steffan, 2011). They can and do use new technology in their own ways, skipping the outdated methods to embrace new ones that fit their own situation. This is called "leapfrogging."

An example of leapfrogging is the use of mobile phones in Kenya. In 2000 there were 15,000 mobile phones in use; by 2011 there were 6 million (Steffan, 2011). Kenyans use their phones to vote. The results are sent by cell phone and the ballots are used for confirmation. There is less likelihood of election tampering. Phones are also used to determine the best prices for crops: farmers find potential buyers by phone, changing the potential place where the farmers transactions will occur (Steffen, 2011). If an individual does not own a phone, there are kiosks that provide cheap and easy access.

Many Kenyans do not have access to banks, and therefore have no way to pay a bill, make a loan, or give a relative money. To address this, Kenyans accept the transfer of airtime minutes by text messaging. This has become an unofficial, electronic form of a national bank (Steffen, 2011). This shows how people can come together to change their situation. This also requires a sense of community and creativity to use what is available in unexpected and new ways.

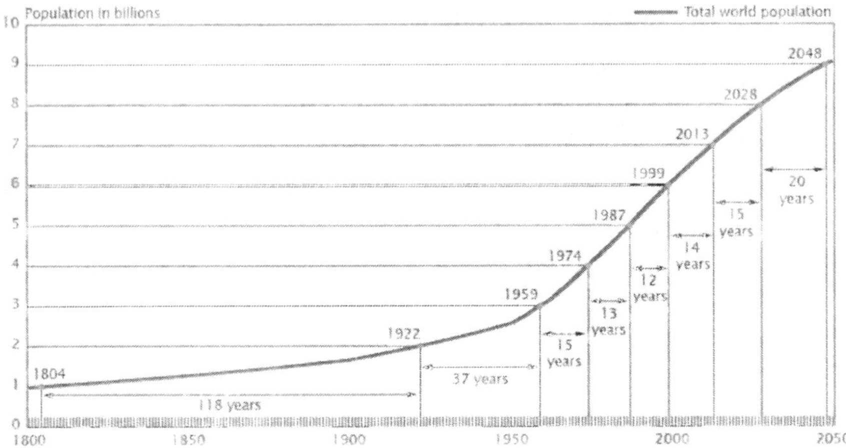

Figure 10.7. Projected increase in world population with 211,000 people per day. Source: United Nations (1995b); U.S. Census Bureau, International Programs Center, International Data Base. Retrieved from https://www.census.gov/population/international/files/wp02/wp-02003.pdf

Wherever we live, the local environment is rooted in our sense of place: our homes, our streets, our neighborhoods, and our communities (Walljasper, 2007). If you want to build a sense of community, the first thing you have to consider is where the boundaries of your community are. You can't instill a sense of community if you don't know what's included within the community and what's not. If your community is to become sustainable, you need to know what is special and what is missing. One place to start would be to map the area you consider to be your community (Steffen, 2011). Don't include too much space or too many people (Walljasper, 2007). How do you decide? It is not easy!

The scale of the traditional, sustainable neighborhood is based on a 5-minute walking radius (1/4 mile or 1,320 feet) (402 m) (Walljasper, 2007). Christopher Alexander (1977) says that a neighborhood should be no more than 300 yards (900 feet or 274 m) across with no more than 400-500 inhabitants, major roads should not be included, and every neighborhood should have an identifiable center (with that number of people, Alexander's neighborhood obviously occurs in a city with 5-story buildings). The city of Minneapolis preferred larger units of 4,000-5,000 people (Walljasper, 2007), similar to the size of Portland's EcoDistricts, perhaps because the population of the cities are 400,000 to 600,000 people. With a city of 500,000, if 5,000 people live in a community, that still translates to a hundred communities, a large number to organize and develop. At the city level, communities have to be large: the city is concerned with the overall structure; the community is concerned about what is happening at the level of the city block.

Bridges and Tangents (n.d.) has a different idea. Communities should be 120-150 people. You should know everyone in your community, perhaps not well, but at least well enough to be able to recognize the members. The average household would have 3 residents, giving this size of community about 40-50 homes or apartments. If the average city block has 10 homes, only 4-5 city blocks are included in the neighborhood. All of these scales are important for varying reasons.

In order to have a viable, sustainable neighborhood, each community needs to have a center and 10 great places to go. As neighbors, we already possess everything we need, plus wisdom and vision, to make our neighborhoods great and unique. Walljasper (2007) says, "It is simpler than you think, more fun than you can imagine, and will improve your life in profound ways." Walljasper has developed a model with 11 principles to guide us called "A Do-It-Yourself Guide to Placemaking" (Figure 10.8). He also describes the importance of offering people a place to sit (a bench) plus the importance of taking back the streets. We have given over our cities to the automobile and so they no longer have the charm and

1. The community is the expert.	7. Form supports function.
2. You are creating a place, not a design.	8. Make the connections.
3. You can't do it alone.	9. Start with petunias.
4. They'll always say, "It can't be done."	10. Money is not the issue.
5. You can see a lot just by observing.	11. You are never finished.
6. Develop a vision.	Plus: Every neighborhood should have ten great places.

Figure 10.8. Adapted from "The 11 Principles" in *The Great Neighborhood Book: A Do-It-Yourself Guide to Placemaking* by Jay Walljasper, 2007. Gabriola Island, BC, Canada: New Society Publishers

individuality they once expressed. Places we love to visit on vacation still hold onto these qualities. At one time in San Francisco, parks were so desperately needed in the city, that an organization called Rebar Art and Design Studio leased a parking spot for only two hours for a "public recreational activity" and rolled out grass, a park bench and a container tree. The popular use of the space over the two hours clearly showed the need for open space in the city (Walljasper, 2007; Rebar, 2005). This activity has turned into a worldwide annual activity called Park(ing) Day that usually takes place in September. In 2005, a single parking space was involved. By 2011, the number had grown to 975 "parks" in 162 cities, 35 countries, across 6 continents. If you wish your city or neighborhood to participate in the Park(ing) Day event, current instructions can be downloaded by visiting Park(ing) Day's website (http://parkingday.org/).

Portland, Oregon, has a unique way of bringing neighbors together: the Share-It Square project, also known as the Intersection Repair or City Repair projects. People from all over the world converge in Portland every year to participate in the event, helping neighborhoods plan and paint intersections, which results in drawing communities together. After the installation, 85% of residents feel that the community has improved with less crime, slower traffic, and better communication within the neighborhood (Walljasper, 2007). All that is required to implement this kind of intersection is the consent of 80% of the residents within a 2-block radius (Figure 10.9).

To bring neighborhoods together in Berkeley, California, and change the growth pattern from city sprawl to high-density, more authentic communities, Richard Register (1987) had a practical vision. If properties were rezoned for open space as they were offered for sale, and downtown areas were encouraged to increase density, sprawl could be not only prevented but also reversed (Figure 8.5). Taking into account the original natural ecosystems, his vision was unique and beautiful. It's too bad that the ideas of the author were not given the attention they deserved when the book was published in 1987.

Cities and towns put rules and laws into place to protect aspects of their residents' lives. However, many of these rules are no longer effective, and in the case of sustainability, stand in the way of progress. Walljasper (2007) gives us some options when it comes to City Hall. You can fight City Hall and you can win. You can work with City Hall. Actually, good government depends on the participation of its citizens. What are you waiting for? You can also remind City Hall that they work for you. Or, you can just forget City Hall. There are some times when it is wiser to act and not ask

Figure 10.9. Share-It Square brings neighborhood residents together. Source: Linda Pope.

permission. And variances can often be obtained for rules that are in the way of real progress (Walljasper, 2007). Do what will make your neighborhood great and strong.

Are there times when we need to consider civil disobedience? There may be times when *not* following the rules is necessary. There are many manuals for anyone looking to change public policy using nonviolent methods. Rinku Sen, the author of *Stir It Up: Lessons in Community Organizing and Advocacy,* helps citizens organize gatherings. What about the Occupy Movement of 2011? Was it a step in community development toward living sustainably and fairly with each other?

How strong is your community? Walljasper (2007) asks the following: In the last week, how often have you done the following?

1. Chatted with clerks, neighbors, people waiting at a bus stop?
2. Said "Hi" to people in passing?
3. Shared a meal, coffee, or a meeting with someone in your community?
4. If you were ill, how many people would bring you meals?
5. How many people could you drop in on uninvited?
6. With how many people could you discuss personal issues?
7. How many people know your goals and support your achieving them?
8. Do you feel you are part of a caring community?

(Walljasper, 2007)

To understand the strength of our communities, a website has formed to provide an overall first-look at our neighborhoods: WalkScore.com®. The site assigns value to neighborhoods (out of 100 points), and makes possible a general comparison between communities, within the same city or between cities across different states (WalkScore®, 2009) (Figure 10.10). Another way is to just ask your neighbors! What would they like to see improved? How should vacant lots be used? Should the community work with the city to purchase the lot for use as a park? What businesses would they like to see move into the empty stores? What is missing from the neighborhood? What would make your community totally unique in your city?

Seasonal activities provide events that draw others to visit your community. In Paris, France, they turn a section of the city into a "beach." The plaza in front of City Hall, the riverfront, and the highway is turned into a beach for the month of July. Sand, deck chairs, palm trees, games and activities all take over this section of the city providing an escape from the city without having to go anywhere at all (Walljasper, 2007)! Is there a place in your community for such a crazy activity?

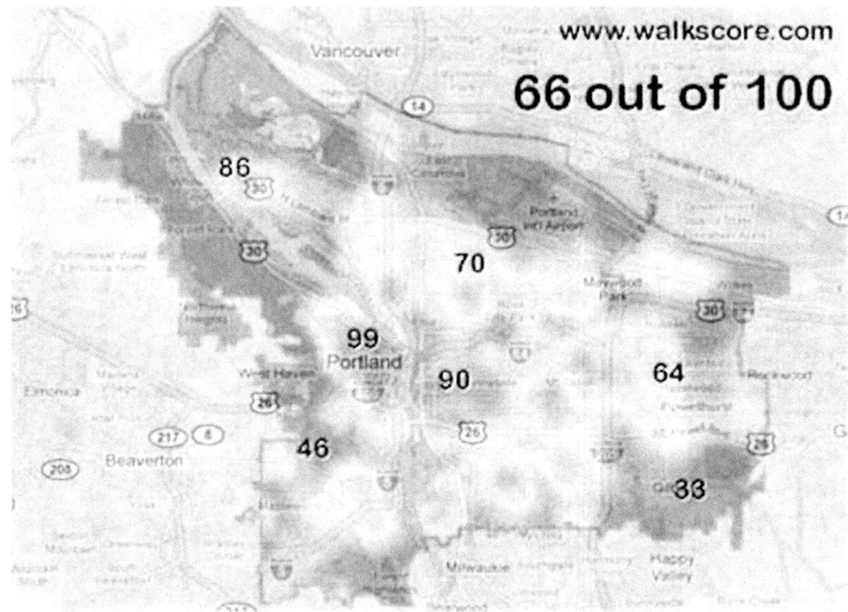

Figure 10.10. WalkScore® results for the city of Portland.
Source: http://www.walkscore.com/OR/Portland

How can you stir up some hope? Walljasper (2007) says two simple things are needed. "One of the most powerful things I learned was that when you transform your immediate environment, your life begins to change" (Walljasper, 2007). If you make your neighborhood better, your life will change. And "Beauty -- no matter how small it is, just a few flowers, is what matters most" (Walljasper, 2007). Change might be just a basket of petunias (Figure 10.11).

E. B. White, an American essayist and author of *Charlotte's Web,* said, "I arise in the morning torn between a desire to improve (or save) the world and a desire to enjoy (or savor) the world. This makes it hard to plan the day" (Shenker, 1969). Perhaps our time should be divided between the two. Is today a day to plant petunias, or a day to enjoy them?

Figure 10.11. The first step towards change in your neighborhood maybe as simple as a basket of flowers. Source: Linda Pope.

References Cited for Chapter 10
People, Culture and Community

Alexander, Christopher. (1977). *A Pattern Language: Towns, Buildings, Construction (Center for Environmental Structure Series. Book 2.* Berkeley, CA: Oxford University Press.

Wang, Fr.Stephen. (2011). "The Perfect Size for a Community is…." *Bridges and Tangents*. Retrieved from http://bridgesandtangents.wordpress.com/2011/05/06/the-perfect-size-for-a-community-is/

Cann, Oliver. (2012). *Slow Progress in Closing Global Economic Gender Gap, New Major Study Finds.* Retrieved from http://www.weforum.org/news/slow-progress-closing-global-economic-gender-gap-new-major-study-finds

City Repair. (n.d.). Retrieved from http://cityrepair.org/about/how-to/placemaking/intersectionrepair/

Doppelt, Bob. (2003). *Leading Change Toward Sustainability: A Change-Management Guide for Business, Government and Civil Society*. Sheffield, UK: Greenleaf Publishing Limited.

Hawken, Paul, Amory Lovins, and L. Hunter Lovins. (1999). *Natural Capitalism*. New York, NT: Little Brown and Company.

Lao-Tse. (6th Century BCE). World Prayers. Retrieved from http://www.worldprayers.org/archive/prayers/meditations/if_there_is_to_be_peace.html

Moore, Steven. (2007). Alternative Routes to the Sustainable City: Austin, Curitiba, and Frankfurt. Lanham, MD: Lexington Books (Rowman & Littlefield).

Rebar Group. (2005). Retrieved from http://rebargroup.org/parking/
Register, Richard. (1987). *EcoCity Berkeley: Building Cities for a Healthy Future*. Berkeley, CA: North Atlantic Books.

Sen, Rinku. (2003). *Stir It Up: Lessons in Community Organizing and Advocacy*. San Francisco, CA: John Wiley and Sons.

Shenker, Israel. (1969). E. B. White: Notes and Comment by Author. New York Times. Retrieved from http://www.nytimes.com/books/97/08/03/lifetimes/white-notes.html

SHRM Survey Program. (2003). 2003 SHRM/NOWCC/CED Older workers survey. Alexandria, Virginia: SHRM Research. Retrieved from http://askjan.org/media/aging.html

Steffen, Alex (editor). (2011). *World Changing: A User's Guide for the 21st Century*. New York, NY: Abrams Books.

WalkScore®. (2009). Retrieved from http://pdxplanningcommissioner.com/2009/12/03/walkscore-and-the-5-10-20-minute-neighborhood/

CHAPTER 11
TRANSPORTATION

Do you (or would you) travel a long distance to get a bigger house, a higher salary, or go to a better school? The problems caused by our cars have more to do with where we live than what kind of car we drive (Steffen, 2011).

You know you are in a stagnant industry when the big innovation in car design in the last decade is more cup holders. After 100 years of engineering, the contemporary automobile is embarrassingly inefficient (Hawken et al., 1999). When Henry Ford designed the first cars, his main interest was that the cars needed to be mass-producible and affordable. The cars included two major design flaws however. Vehicles are 20 times heavier than the driver, and the engine is 10 times larger than is needed for average driving conditions (Hawken et al., 1999). When we begin to shape cars more like aircraft and less like tanks, magical things will start to happen.

Fifteen percent of the cost of a steel car pays for all the steel. The remainder is the cost of pounding, welding, finishing that steel. The average car has 200-400 parts and each part requires 4 dies to make that part. Each die costs a million dollars. Painting is the costliest, most difficult, most polluting step in the production. The design requires thousands of engineers and billions of dollars worth of steel dies, and it takes decades to recover the costs (Hawken et al., 1999).

Of the power required to move a car forward, one third is used to accelerate the car (heating up the brakes when you need to stop the car). Another third of the power heats up the 6-7 tons (5.4 t to 6.4 t) of air as the air is pushed aside when the car moves forward. The last third of the power heats up the tires and the road due to rolling resistance. All of these power losses need to be reduced. If these three key elements are redesigned, cars can save 70% to 80% of their fuel consumption while making them safer, sportier and more comfortable (Hawken et al., 1999).

The first goal should be to make the cars ultralight (2 to 3 times less heavy than steel). The design should provide ultra-low drag to avoid air resistance and to roll easier. If these considerations are implemented, less power will be needed so the propulsion system can easily be a hybrid or electric (Hawken et al., 1999). With these designs incorporated, the car could be more durable and cost less. The new vehicle could "blend the comfort and refinement of Lexus, the stiffness of Mercedes, safety of Volvo, BMW acceleration, Taurus price, four-to eightfold improved fuel economy (80-200 mph)[129 to 321 kph], 600-800 mile [966 to 1,287 km] range between refueling, and zero emissions" (Hawken et al., 1999). All of the technology already exists.

If carbon-fiber composites are used for the autobodies, breaking and rolling resistance are reduced due to the two- to threefold weight reduction. Streamlining the design can cut 40% to 60% of the air resistance. Double-efficiency tires already on the market cut rolling resistance 66% to 80%. If these changes were implemented, power needs would be cut in half. Ten to 20 years are required to implement changes. Hawken et al. wrote about this in 1999; 15 years have passed. What are we waiting for? Where is this new and better car?

The average American car is parked 96% of the time. In the future, leasing cars (as well as everything else!) will be the normal method of obtaining a new car. While your car is parked, others will have the opportunity to use the car to run errands: car sharing. A new company, Getaround, is the beginning of this new format (Getaround, 2014). Also, instead of plugging into the grid to recharge your batteries, your car would be generating electricity (more than 20 kilowatts) and sending it into the grid to pay a large portion of the lease (Hawken et al., 1999). It is all a matter of perspective. What is the purpose of having a car? Is it something we will need in the future or will there be other options?

The Rocky Mountain Institute is generating emerging technologies that will be less harmful to the environment. Their Hypercar incorporates some of these technologies. When the car was completed, the company put its design into public domain (making it un-patentable) in order to trigger a design revolution. The newly designed composites emerged from the mold already shaped and finished (Hawken et al., 1999). The single, low pressure step reduces the expensive traditional tooling costs associated with a steel car by 90%; space and effort required for assembly is also reduced 90% because parts are so lightweight they do not require a hoist to be lifted. Painting is not needed because the color is incorporated into the molding process. The design teams can be small; production runs can be low; the break-even point is also low; and the potential for design modifications is enhanced (Hawken et al., 1999). The Hypercar is the fastest car in the world (285 miles per hour or 459 kph!). With all this flexibility, why do cars have to come from Detroit? Or Japan? Why not your garage? We need more local manufacturers. What should an Oregon car look like? Would a Willamette Valley car be different from a high-desert car? A family car versus one for teenagers?

How can we encourage this transformation? At the state level, we could consider a “feebate.” This would mean that when you buy a new car, you either pay a fee or get a rebate, depending on the efficiency of the new vehicle. Or you get a rebate on the new car depending on how much more efficient the new car is, as opposed to the one being scrapped (Hawken et al., 1999).

Personal vehicles and parking problems go hand in hand. Affordable housing is hard to find because we currently require one or two free parking spaces to be included in the rent or purchase price. This practice may be changing. In San Jose, California, developers are not allowed to provide a free parking space, but must include free transit passes with each unit. In Frankfort, Germany, office workers are required to purchase the parking space for the car (Hawken et al., 1999). Companies are required to pay higher taxes in Britain if they provide free parking for their employees. In Tokyo, it is not possible to purchase a car unless you can prove that you have rented or purchased a place to park your car (Hawken et al., 1999). We take parking for granted in the United States.

Besides parking, we also have to deal with congestion. People in Bangkok have some of the worst traffic problems, spending a full 44 days a year in gridlock. Singapore has avoided this issue by charging \$3 to enter the city. This one action reduced the number of cars on the road by 44% and the number of single-occupancy cars by 60% (Steffen, 2011). London followed suit, but charges \$14 to enter the city core. This zone is monitored by camera. If you do not pay in advance, you are charged an even higher rate, and if you try to avoid paying, that will result in a fine of $267. This has caused 30% fewer cars entering the city center. Other European capitals have similar plans that have also decreased the traffic loads (Steffen, 2011). Actually, in many European cities, 40% to 50% of trips are taken by walking or biking, and only 10% by transit (Hawken et al., 1999). It is much better if you can arrange your life so that you work where you live.

In Curitiba, Brazil, a city designed around their transit system, the bus comes every minute during rush hour (Hawken et al., 1999). Do you stop and pick up a couple groceries on the way home from work if you take the bus? In Curitiba, why not? The next bus is only a minute away. Also, you pay to get into a "glass tube" rather than waiting in a single slow-moving line to pay as you get onto the bus (Figure 10.5). This means everyone can get off and on faster; you stay out of the weather at the same time. Obviously, this is a superior system.

In America in 1990, 54% of the population lived within 5 miles of work, although only 3% biked and even less walked. Only 3% of our trips are by public transit; 87% are by car. If we could shift only 5% of non-rush-hour miles to bikes, the total social savings would be more than 100 billion dollars (Hawken et al., 1999).

There are many ways we can climb out of our mental boxes regarding cars. One college student leased his car to a taxi company while he was in class. This earned him enough to buy a new car every 2 years plus it paid for his education (Hawken et al., 1999). Getaround Portland (n.d.) is a new organization working to make Portland the car-sharing capital of the world. You can share your car with local Portlanders and earn thousands of dollars at the same time (Getaround, n.d.). Want to retrofit your car to be electric so that you have zero emissions? Many are doing just that across the country (Electric Car Conversion, n.d.). Of course the source of the electricity must be kept clean as well.

More than half of Americans could be working from home. It would cost less to supply fiber optics to every home in American than it costs to rebuild our roads every 2 years (Hawken et al., 1999). This would remove millions of cars from our roads, as well as eliminate the emissions they would cause.

Portland, Oregon, determined in the 1970s that gasoline consumption would be 5% less if the city resuscitated the presence of the neighborhood grocery store. The previous CEO of New Seasons Market, Lisa Sedlar, has started a new neighborhood grocery store chain called Green Zebra (named after a tomato that does well in the Portland climate) (Friedman, 2013). Although much smaller than other local grocery stores, I feel that they could go even smaller. What are the 10 items most purchased in the “cashier-yourself-out” lines? Those items should be available in small corner grocery stores that are within walking distance. The 7-Eleven and Plaid Pantry convenience stores are already on every corner. They have the potential to be so much more than just over-priced beverages, cigarettes, and candy.

All of this points again to having better neighborhoods. We don’t really need better transit, we just need better neighborhoods. If we could live, work, shop, and play within an easy 5-minute walk, why would we need cars or buses? What is missing from your neighborhood that prevents you from living within a 5-minute walk to everything?

The people of Amsterdam, Holland, were thinking in this direction and took a different approach. They gradually banned cars from their city by slowly making the sidewalks wider and by adding more bicycle lanes. Parking spaces are difficult to find and very expensive. They also set the speed limit at 18 mph (29 kph) (Hawken et al., 1999)! You know that 20 mph (32 kph) school zone that you hate to drive through? In Amsterdam you could get a speeding ticket if you drove that fast! Driving at 18 mph (29 kph), it is much more practical to leave the car home and ride a bike.

Copenhagen, Denmark, has the world's most impressive bike culture. They have bicycle lanes wide enough for two to ride side-by-side allowing conversation between bikers. Adjacent to this wide lane is a skinny lane for those cyclists who have a timetable to keep and need to keep moving (Steffen, 2011). The city has also developed the "Greenwave" system, which times stoplights so that cyclists never hit a red light. That means the lights are timed for 12 mph (19 kph), another way of making the choice of bikes better than that of cars.

Tokyo has designed automatic underground bicycle parking garages. An Eco-cycle parking kiosk can hold up to 9,400 bikes and retrieves your bike in a matter of a few seconds (Steffen, 2011).

Do the Portland hills discourage you from even trying to bike them? All that constant change in elevation can really be daunting to a beginner. In Trondeim, Norway, they have the same problem. They, however, have installed a bike-lift system that allows you to remain on your bike. You activate the lift with a keycard, and it pulls you and your bike up the hill (Figure 11.1) (Steffen, 2011).

Portland, Oregon, has the highest share of bicycle commuters (6% to 8%) of any large U.S. city. We are also honored with many unique and individualized creations, from tall bikes to bikes intended to carry both cargo and whole families (Figure 11.2). Many of Portland's festivals incorporate bike owners into the city's special events. The Portland World Naked Bike Ride that occurs the first week of June attracts thousands of bikers every year, young and old. Once a month, Sunday Parkways attracts over 85,000 Portland area residents and visitors to ride their bikes on various 6-10 mile (10-16 km) loops within the city (Sunday Parkways, 2014).

When you think about your community, does it have everything it needs within a 5-minute walk? How well connected is it to the rest of Portland? What would have to change so that you would give up your car? Begin to think outside the box.

Figure 11.1. Bike lifts in Trondheim, Norway, to assist cyclists with difficult topography. Source: Wiki Commons.

Figure 11.2. Portland bikes reflect the creativity of the owners. Source: (left) ItzaFineDay, Flickr.com; (right) Linda Pope.

References Cited for Chapter 11
Transportation

Doppelt, Bob. (2003). *Leading Change Toward Sustainability: A Change-Management Guide for Business. Government and Civil Society*. Sheffield, UK: Greenleaf Publishing Limited.

Electric Car Conversions for Under $500. (n.d.). Retrieved from http://www.youtube.com/watch?v=k62FzVuSF58

Friedman, Nicole. (2013). Green Zebra Grocery, founded by former New Seasons CEO, opens healthy convenience store. *The Oregonian*. Retrieved from http://www.oregonlive.com/business/index.ssf/2013/10/green_zebra_grocery_founded_by.html

Getaround. (n.d.) Retrieved from http://www.getaround.com/portland
Hawken, Paul, Amory Lovins, and L. Hunter Lovins. (1999). *Natural Capitalism*. New York, NY: Little Brown and Company.

Steffen, Alex (editor). (2011). *World Changing: A User's Guide for the 21st Century*. New York, NY: Abrams Books.

Sunday Parkways. (2014). Portland Bureau of Transportation. City of Portland. Retrieved from https://www.portlandoregon.gov/transportation/58929

CHAPTER 12
THINGS WE MAKE AND PURCHASE

The average American strives for a lifestyle that requires an annual income of $100,000. The problem is that the average American income is only $28,000 or, for a family, $49,445 (Steffen, 2011). This means that we spend our lives striving for a life we can never achieve, or we go into serious debt trying to get there.

Due to the inventions of the spinning jenny and spinning mill, a single worker in 1812 could do the work previously performed by 200 spinners in 1770 (Hawken et al., 1999). With further innovations over the next 5 decades, textile production increased 120 times over (Hawken et al., 1999). At the beginning of the industrial revolution, there were few workers, and they were overworked. Resources seemed limitless. Today the situation has been reversed (Hawken et al., 1999). We have many people and few resources.

Industry uses over a million pounds of materials per person per year, most of it wasted or invisible (Hawken et al., 1999). We found one way to make something, but it was just that, only one way. There are a dozen other ways to make the same item using fewer resources. If we want a prosperous future, we have to recognize the limits of our resources, and make them more productive. This involves getting 4, 10 or even 100 times more from each unit of energy, water, materials, or anything that we take from the planet. A way to more fully understand the issue is to look at the "backstory" of everything we purchase.

What happened to our "stuff" before we bought it? Who made it? Where was it grown or mined or manufactured? Did the farmer use fertilizers or pesticides, or integrated pest management? Antibiotics or free-range grazing? How did the item get to you? How were the profits distributed? Consumers are ready to hear backstories (Steffen, 2011). A list of all the materials and manufacturing processes required for a laptop computer is available at Sourcemap (n.d.) and includes a video that fully explains the backstory concept. In another Sourcemap example, 18 breweries in northern Scotland were shipping their product over 4,000 miles (6,500 km) one-way for bottling (Figure 12.1, left). A new plant would reduce the shipping

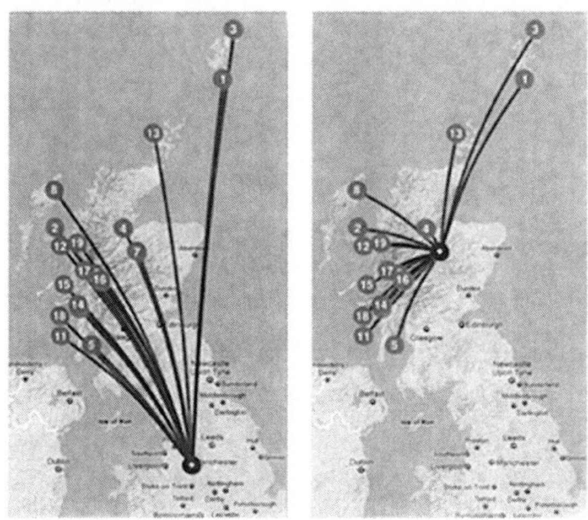

Figure 12.1. Sourcemap for the production of beer bottling in Scotland. Source: Bonanni et al (2010).

distance for bottling by two-thirds (Figure 12.1, right). An additional question still needs to be asked: where is the beer sold or consumed? London? Or in northern Scotland? Where would emissions and wastes for bottling be reduced most: in the north or south?

We are just beginning to enter an era of holistic accountability and transparency. Companies are being held responsible for the backstory of their products (Steffen, 2011). Consumers are asking questions and want to know everything that happens to get the product or service to them. And they want to know what will happen after it leaves their lives.

It is hard to sort out what labels can be trusted. An organization called EcoLabel Index is the largest global directory that tracks labels in 197 countries and 25 industries. Christien Meindertsma (2008) traced all of the products derived from a single pig in the book *Pig 05049*. You can watch the author show you all the pages in the book on Vimeo (Meindertsma, 2008). He shows us how little we know about the products around us. Where do they come from? And where do they go?

Why do we ship our chickens to China to be slaughtered and flash-frozen, and then shipped back to us? With all the problems we are having with meat processing and salmonella food poisoning, it would seem that we should be more concerned and be more willing to take the situation into our own hands, i.e. produce some of our own meat, or at least buy from local butchers and farms.

But it is not just food choices that need new consideration. The common shoe is actually hazardous waste. Every step you take on a rough surface or on a sidewalk abrades the shoe sole, creating dust to be inhaled. Chromium used in the leather tanning processes causes damage to the respiratory system. In one shoe factory, all the workers were elderly and wore gas masks in the factory. The supervisor informed the authors (McDonough and Braungart, 2002) that it took an average of 20 years for workers to develop cancer from the chromium exposure so the company decided to hire only workers older than 50 years of age to work with the dangerous substance. Do you think that is an appropriate response to the situation? Or should other chemicals be considered in the tanning process? James and Lahti (2004) describe a natural process that takes months (instead of hours) to tan leather, uses only water and spruce bark, and uses no toxic chemicals.

What should we do with these shoes when we are through with them? A study by Loughborough University in Great Britain clearly describes the difficulties with our current manufacturing processes and the potential recycled products that are possible with different processes:

- Recovered leather fibers can be re-formed to produce bonded leather sheets;
- Reclaimed rubber can be used as a running track or playground surfacing product;
- For some types of footwear rubbers, finely-ground rubber can be put back into new shoe soles, achieving so-called “closed loop” recycling;
- Recycled foams can be used in underlay material for laminate floors and carpets; and
- Mixed textiles and other lighter residues could be used as insulation material for buildings. (Green Guide UK)

High-tech products are frequently made of low-quality materials. Banned in the U.S. and Europe, the carcinogen benzene is in rubber products made in developing countries that have not banned the chemical. One of these rubber products becomes a part that is assembled into a treadmill and that emits the carcinogen as you exercise (McDonough and Braungart, 2002).

The low-tech products designed for our children are not exempt from exposing them to hazardous chemicals. A child's skin is 10 times thinner than the skin of an adult. Kids' plastic swimming pools contain plasticizing phthalates. You may say, "I don't know of any kids that have gotten sick from a plastic float or swimming pool." But many kids develop allergies, multiple-chemical sensitivity syndrome, asthma, or they feel unwell without knowing why (McDonough and Braungart, 2002). Becoming aware of the chemicals that can harm us is a first step to eliminating them from our environment.

Janine Benyus (1997), a biologist, is causing us to become aware of how we manufacture things in comparison to how things are manufactured by nature. We use high temperatures, high pressures, and strong chemical treatments, or "Heat, beat, and treat" (Benyus, 1997).

Nature doesn't follow that path. Nature manufactures under life-friendly conditions—in water, at room temperature, and without harsh chemicals or high pressures (Benyus, 1997). There is no difference between what nature needs to do and what we need to do. "Life manufactures, computes, does chemistry, builds structures, designs systems, and engineers (to within a fine tolerance) the tools needed to fly, burrow, build dams, heat or cool homes, etc." (Benyus, 1997).

For example, what can we learn from the horn of the rhinoceros? It actually contains no living cells. But when a crack appears, it quickly patches itself. The material disassembles, fills the crack, and then reassembles again. This is a key to designing products that never need repairs. Imagine an airplane being hit by a flock of birds, and instead of needing to immediately land, the plane could heal its damage in-flight.

Following the biomimicry principles proposed by Benyus, in the future we will not ask, “Is it good for us?” or “Is it profitable?” Instead we will ask:

- Does it run on sunlight?
- Does it use only the energy it needs?
- Does it recycle everything?
- Does it reward cooperation?
- Does it bank on diversity?
- Does it utilize local expertise?
- Does it curb excess from within?
- Does it tap the power of limits?
- Is it beautiful? (Benyus, 1997)

Incorporating these ideas, designs, and purchases for the future will recover the ideal of making things for everyone that will last for generations. Purchases should be something that we want to keep forever. For example, computer or cell phone cases should be unique and exquisite, with only the contents being replaceable or upgradeable (Steffen, 2011) (Figure 12.2).

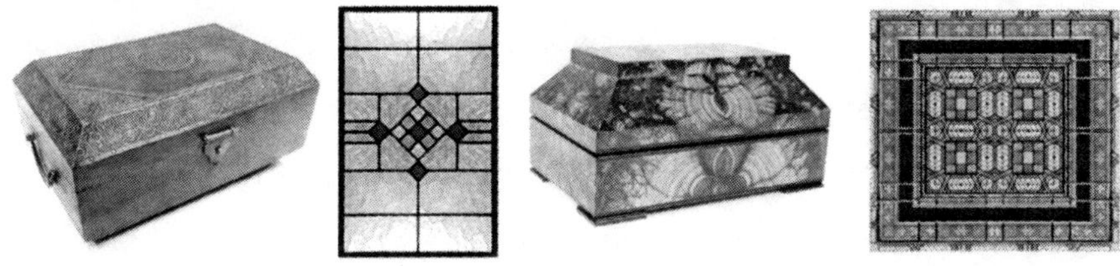

Figure 12.2. Variety of unique containers that could be computer cases designed to last for generations. Source: Bigstock.com; iStock.com.

In the process of changing the way that we make things, there will be surprises along the way. German automobile manufacturers developed the extended producer-responsibility and product-takeback plan, a lifetime responsibility for the manufacturer to recycle its own products. At first it was a real headache. Then the process started to save the company money. It turns out that cars that are made so that they are easy to disassemble are also easier to assemble and repair (Steffen, 2011).

In the course of doing experiments, the average chemistry class turns $8,000 worth of pure, simple reagents, intended to elicit a specific chemical reaction, into a complex soup of nasty toxins that cost $16,000 to discard properly. A University of Zürich professor decided to reverse the process, and designed experiments that turned the toxic wastes back into pure, simple reagents. The wastes for the program were reduced 99% and $20,000 per year were saved (Hawken et al., 1999). It was a matter of developing a new perspective.

Design Tex makes office furniture. Environmentally, fabric trimmings are defined as toxic waste (Hawken et al., 1999). Design Tex decided that they wanted to mulch their fabric trimmings! In order to do that the fabric had to be "free of mutagens, carcinogens, heavy metals, endocrine disruptors, persistent toxic substances, and bio-accumulative substances" (Hawken et al., 1999). They approached 60 companies with the task; all but one declined, Ciba-Geigy. Of the 8,000 chemicals that they previously used, 7,962 were eliminated. Only 38 chemicals were kept and the new line was designed around those chemicals. As a consequence, the water leaving the factory was as clean as the water going in (Hawken et al., 1999).

What cleaning products do you use? Seventh Generation is known for its environmentally friendly products. Their sales are over

100 million dollars; they employ less than 100 employees; and their sales in 2008 grew 50%, leading in every category. At the same time, they reduced greenhouse gas emissions 34%, with a goal to reduce them 80% by 2050. The company has corporate giving (gifts to charity) at a rate of 10% of their pretax profits.

What do we need in our cleaning products? First, products should be concentrated, with water reduced to a minimum. No need to expend the energy to ship water from place to place. Actually, there should be cleaning properties designed for each locality. Is the water hard or soft? Are washing machines used or do local citizens wash by pounding on rocks? In the laundry cycle, only 5% of detergent is consumed (McDonough and Braungart, 2002). Can we redesign washing machines to recover detergent and re-use it multiple times?

We can move beyond asking whether a detergent is gentle on our hands to asking if it's gentle on rivers or lakes. Once you really understand what kind of soap the customers want, you can figure out what kind the river wants. Learning from nature (biomimicry) will give us new insight. What can we learn from the petal of a flower?

Figure 12.3. Flower and potential texture of a petal. Source: (left) L. Pope; (right) Bigstock.com.

The texture of the flower petal is a field of small bumps on which dirt or debris rests, not adhering to the petal itself (Benyus, 1997) (Figure 12.3.). When it rains, the water rinses dirt off the petal easily. Some manufacturing companies are searching for ways to add this texture to the fabric used in our clothing so that only water would be necessary to remove dirt from our clothing, and no soap would pollute our environment.

Alternatively, we will switch to a service economy, and will pay a monthly fee to have clothes cleaned. Washers will have a counter on them, like copy machines. The manufacturer will retain ownership and responsibility of maintaining the machines, replacing them with newer models when appropriate, repairing as needed for no additional charge. The same would apply to cars, DVRs, computers, refrigerators — more examples of cradle-to-cradle manufacturing (Hawken et al., 1999).

Imagine how things would change if the only physical objects you bought were those you truly wanted to own for either sentimental or aesthetic reasons. Everything else would be leased, with providers installing, maintaining, upgrading, and eventually replacing your appliances, furniture, even your cookware (Benyus, 1997).

The cradle-to-cradle process will change the manufacturers' viewpoint, creating quality instead of planned obsolescence. A new level of engineering will be required. World master engineer Eng Lock Lee is such a person. He makes the same products, uses the same hardware and handbooks, but his designs are 3 to 10 times more efficient and they cost less to build (Hawken et al., 1999). In his designs, energy is used frugally, recaptured, and reused until nothing is left. Lee says, "You know you are on the right track when your solution accidentally solves several others" (Hawken et al., 1999).

As consumers, we need to question all aspects of our purchases and to think about each consequence. In our clothing, for example, most natural fibers are not grown sustainably. Cotton uses 25% of all agrochemicals and insecticides, and requires excessive water (1 lb of cotton fiber uses 2.5 tons of water) (Hawken et al., 1999).

Although many of us prefer cotton clothing, synthetic fibers based on petrochemicals have postponed the clearing of huge acreages of trees for cotton planting. The fabrication of synthetic fabrics continues to improve. A 300-acre (121 ha) petrochemical plant and a small natural gas facility can match the fiber production of 600,000 acres (242,811 ha) of cotton (Hawken et al., 1999). However, some of us choose not to turn to synthetic fibers because of the fossil fuels that are required.

What clothing should we purchase? Cotton is problematic due to the heavy load of chemicals and the amount of water used in its growth. If we all decided that we had to have natural-fiber jeans dyed with natural dyes, millions of acres of cropland would have to be converted to cultivate indigo and cotton, acreage otherwise needed for food. The natural dye indigo, grown in Nigeria, contains mutagens and, grown in monoculture, decreases local diversity (McDonough and Braungart, 2002). Also, we cannot grow cotton in Oregon.

What if we had the limits of the locavore (200 miles [322 km]) for our clothing purchases? Where does most of our clothing come from? Check the clothes you are wearing right now. Where did they come from? What is *really* made in Oregon? If we cannot grow cotton, what can we grow? We have wool from Pendleton. What are our other choices? Hemp? Bamboo? What are the manufacturing options that we can start now, in Oregon? What are the options where you live?

In the design of other types of products, computer software can help, using the least amount of material possible and making the product as strong as it needs to be, but no stronger. This process requires less material because strength is put only where needed (Hawken et al., 1999). For example, a toilet float-valve assembly has changed from 14 parts costing $3.68 to one part costing $0.58. A 13 lb (5.9 kg) tricycle made of 126 parts was redesigned to a 3 lb (1.4 kg), 26-part version at 25% of the original cost. A windshield-wiper arm was reengineered from 49 parts to one, and at a lower cost even though it now uses a $68 per-pound ($150 per kg) carbon fiber composite (Hawken et al. 1999). This redesign process is called "dematerialization." Using this process, materials required to make a yogurt container in Europe dropped 67% over the period 1960-1990. From 1970 to 1990, the amount of glass used in a beer bottle dropped 28%. Office buildings needed 100,000 tons (90,718 t) of steel 30 years ago; now they use 35,000 tons (31,751 t) because of better steel and smarter design (Hawken et al., 1999).

In Japan, parts suppliers are connected by computer, making only what the factory needs on an hour-by-hour basis. This means fewer materials need to be on hand as fewer parts, etc., need storage space, and less overproduction occurs (Benyus, 1997). This process is called "Just-in-Time," or JIT, and works very well in the manufacturing environment. However, since the 2011 earthquake in

Japan, JIT is being re-thought, as the country ran out of food in just 2 hours due to using JIT methods in food processing and stocking.

In France, 10 million buildings receive heat through contractors called chauffagistes, who work in an industry comprised of 160 firms, employing 28,000 people (Hawken et al., 1999; Ruppert, 2009). These companies do not sell energy in the form of gas, or oil, or electricity. Customers do not want oil, etc., they want warmth! The chauffagiste is contracted to supply warmth (a specific temperature range) to a specified floorspace (apartment or home) for a certain cost. It is up to the chauffagiste as to how this is accomplished. He may provide a more efficient heating system, or insulate the building. The less energy used, the fewer but higher quality materials involved, the more money the chauffagiste earns (Hawken et al., 1999; Ruppert, 2009).

Though previously competitive with one another, American automobile manufacturers have begun to cooperate to form the Vehicle Recycling Partnership (VRP) (Benyus, 1997). General Motors, Chrysler, and Ford are working together to develop common labeling and materials so that they are able to reuse each other's parts. When the automobile is dismantled, some parts will be reused, others recycled, and still others processed into new parts. VRP and the United Council for Automotive Research (USCAR) have a mission statement that aims to "promote an integrated and sustainable approach to improving the technical and economic feasibility of vehicle recycling in North America" (2013).

It is a challenge to find products that have been made in a sustainable manner. The Good Guide (2013) has evaluated over 100,000 products in the areas of nutrition, scientifically proven hazards, animal welfare, recycled materials, pollution, fair trade, organic, fragrance-free, energy efficiency, climate change, resource conservation, labor, and human rights. The Good Guide also has a mobile phone application that can be personalized to make shopping easier.

Purchases made by local residents in other countries will vary greatly. For example, in India, the Honey Bee Network collects information from isolated communities in order to spread traditional knowledge faster (Steffren, 2011). Who makes a bicycle hoe? Where can I get a micro-windmill battery charger for my cell phone, or a pedal-operated washing machine?

Overall, how we make our products in the future will be directed more by what will happen to them when we are through with the item. How is this item going to be reused or recycled? Can the materials come from a local source?

When Ray Anderson started the process of producing “green” carpets, he soon realized that it meant his company and all its suppliers also had to be green. There is no such thing as a “green” product coming from a “brown" company (Anderson, 2006). At some point, current manufacturers and designers have to decide. And some will say, “We can’t keep doing this. We can’t keep supporting and maintaining this system. When is that point? We say that point is today, and negligence starts tomorrow” (McDonough and Braungart, 2002, p.43). The same idea applies to the individual. Is today the day you will decide to begin a new sustainable path?

References Cited for Chapter 12
Things we make and purchase

Anderson, Ray. (2006). Restoring the Planet, One Carpet Tile at a Time. Green Technology. Retrieved from http://www.green-technology.org/green_technology_magazine/ray_anderson.htm

Benyus, Janine. (1997). *Biomimicry: Innovation Inspired by Nature*. New York, NY: HarperCollins Publishers Inc.

Bonanni, Leonardo et al. (2010). "Small business applications of sourcemap." Proceedings of the 28th international conference on Human factors in computing systems—CHI '10. Atlanta, Georgia, USA, 2010. 937. c2010 ACM. Retrieved from dspace.mit.edu/openaccess-disseminate/1721.1/60957

Good Guide. (2013). Video available on YouTube at: http://www.youtube.com/watch?v=rlT6kqHFjvU Retrieved from http://www.goodguide.com/#

Green Guide Footwear Recycling. (2013). *Reducing landfill, creating recycled materials: Loughborough University develops a new footwear recycling system*. Retrieved from http://www.greenguide.co.uk/node/4605

Hawken, Paul, Amory Lovins, and L. Hunter Lovins. (1999). *Natural Capitalism*. New York, NY: Little Brown and Company.

James, Sarah and Torbjörn Lahti. (2004). *The Natural Step for Communities: How Cities and Towns Can Change to Sustainable Practices*. Gabriola Island, B.C., Canada: New Society Publishers.

McDonough, William and Michael Braungart. (2002). *Cradle to Cradle: Remaking the Way We Make Things*. New York, NY: North Point Press.

Meindertsma, Christien. (2008). *PIG 05049*. Flocks; Second edition and YouTube at http://vimeo.com/10717795

Ruppert, Jeff. (2009). *LEED Driven Services – The Chauffagiste*. Retrieved from http://buildearth.org/?p=445

Sourcemap. (n.d.) Sourcemap for a Laptop Computer. Retrieved from http://free.sourcemap.com/view/744 and video from http://vimeo.com/36053523

Steffen, Alex. (2011). *World Changing: A User's Guide for the 21st Century.* Abrams, NY: Sagmeister Inc.

USCAR. (2013). United Council for Automotive Research LLC. Retrieved from http://www.uscar.org/guest/publications.php

CHAPTER 13
WORK

Many of us dream of having the perfect job, maybe working from home. We want to be self-sufficient, "off-the-grid," clothes-optional (according to one of my students), but still be on the Internet. There are some things we just don't want to live without, now that we have experienced them.

We all have to work, but is our work environment any better than in other parts of the world? Europeans have longer vacations and shorter working hours. People who live in Western Europe are far healthier than Americans, and are less likely to have chronic illnesses, even cancer (Steffen, 2011). Europeans exercise more, socialize more, sleep more, and all this is made possible because they have more time than Americans. Think about how we structure work.

If we are going to make a difference in the world, in our careers, in our companies, it takes more than just inspiration, and waiting for someone else. We have to integrate sustainability, ecology, and social justice into whatever fields we pursue (Steffen, 2011). In how much of your life does the discussion about sustainability occur? How often does the topic come up at work? If it is not happening as often as you think it should, perhaps you are the one that has to start the conversation.

Start a conversation, you say? I can't rock the boat! It's too hard to find a job. I need to keep that job, even if they do not act in a sustainable manner. OK. So rock your competitor's boat! Rock all the other boats. If others start to change, your company will have to do so as well, if they want to stay competitive. Find whatever ways you can, in your personal life, to get others to change. We have already talked about the waste that occurs everywhere. I am sure that you see it in your workplace, and that you can convince co-workers that waste in any business must be eliminated, to both protect the earth and to improve the bottom line.

At the end of World War II, for every $1 Washington raised in taxes on individuals, it raised $1.50 in taxes on business (Wolff, 2011). In contrast, today, for every $1 Washington gets from individuals, it gets 25 cents in taxes from business. Business has somehow successfully shifted all of the responsibility of the federal tax burden onto individuals (Wolff, 2011). We have to work much harder, often with two or more jobs, to compensate for this imbalance.

What is important in our work? What do we strive for? It varies with culture. In a Buddhist society, spiritual enlightenment is the ultimate goal in everything one does. In capitalism, we want more stuff: a bigger house or car, a higher standard of living, or maybe just more money in the bank. "The American version of capitalism may not be sustainable, since the wealth and the satisfaction of artificial desires, without taking into account the consequences, cannot be met forever, for it defies the laws of nature" (Khanthong, 2009). Think about the growth pattern of yeast! We cannot just consume resources and reproduce forever. In capitalism, we fire employees so that we have 1% more in our bottom line. In contrast, Buddhist-style economics try to see how many people they can hire and still make a profit (Khanthong, 2009).

True wealth is measured by how happy we are. A study issued by the United Nations Human Development Reports says that the number of Americans considering themselves to be happy peaked in 1957, even though the GDP has nearly doubled (Nathaniel, 2011). The ever-expanding economy has no correlation with our well-being (Nathaniel, 2011). In Bhutan, the government is thinking about how economic success ought to be measured. Instead of Gross Domestic Product, they are using Gross National Happiness instead (Sivaraksa, 2009). In order to promote personal development, they focus on "small-scale, indigenous and sustainable alternatives to globalization" to restructure their economics to reflect their Buddhist principles (Sivaraksa, 2009).

In order to increase the number of jobs in America, we need to increase the number of manufacturers. It used to be difficult or impossible to become a local manufacturer because the capital outlay was high. The Internet makes it possible to fabricate your ideas almost immediately with eMachineShop.com. You can put together your own "Fab Lab" by MakerBot (Flaherty, 2009) for about 1,000 dollars. This is a fabrication laboratory of open-source computer-controlled design and manufacturing equipment (Steffen, 2011). If you want to duplicate something the size of an egg or larger, the MakerBot will do a reasonable job. Other 3D printers cost 10 times more (Flaherty, 2009).

What about funding small businesses? If we could increase efficiency in the military budget, we would gain needed funding to create green jobs and green manufacturing. What was the U.S. military budget in 2010? It was $663,255,000,000, that is 663 *billion* dollars. The 2007 Green Jobs Act authorized 125 *million* dollars per year to train workers in renewable energy (Edwards, 2010). The government authorized 2 billion dollars a year for local governments to implement energy and climate change measures. But the cost of a single wind turbine is 1 million to 2.5 million dollars each. How far can those 2 billion dollars go? But if the military budget found efficiencies that could reduce their budget by 10%, there would be an additional 66 *billion* dollars to help with these changes (Edwards, 2010).

Where do you work? Does it have a sustainability plan? What do you see yourself doing in the future? Will you be doing the same work, or something else? Where do you see new opportunities that were not there before you were given a sustainability viewpoint? How can we help change jobs to become more oriented towards sustainability?

Many businesses need help to overcome costs, pricing, or market disadvantages. One example is the business of education. Parents do not pay the full cost of their children's education; it is subsidized and the government pays a portion of the cost (Hawken et al., 1999). Subsidies occur in agriculture, energy, transportation, water, forestry, and fisheries. The sum is enormous, amounting to 1.5 trillion dollars a year. It may be time to rethink the role that these subsidies play (Hawken et al., 1999).

Subsidies leave the environment and economy much worse than before. They inflate the cost of government by adding to deficits and cause taxes to be high. They confuse investors and suppress innovation (Hawken et al., 1999). Subsidies encourage consumption instead of productivity and conservation. They benefit the rich and disadvantage the poor (Hawken et al., 1999). Can this money be better used another way?

It may help to know the backstory of workers' rights, for example. Migrant farm workers make less than $10,000 per year. Tomato pickers in Florida earn $0.45 to pick a 32 lb (14.5 kg) box of tomatoes (Steffen, 2011). Fast food chains have held wages stagnant for three decades (Steffen, 2011; Hightower, 2013). Is this an appropriate way to treat our citizens and employees?

What we need is a practical step toward resource productivity. If we shift taxes away from income, and place them instead on carbon fuels, resource exploitation, pollution, and waste, all of which are currently subsidized, this will encourage more employment. These taxes would also discourage activities that cause social or environmental damage (Hawken et al., 1999). If we had a carbon tax, what would be taxed?

- Gases that cause climatic change;
- Nuclear power;
- Any non-renewably produced energy;
- Air traffic;
- Vehicular use and public roads;
- Car insurance collected at the pump, eliminating government subsidies of uninsured motorists;
- Pesticides, synthetic fertilizers, phosphorous;
- Piped-in water;
- Old growth timber;
- Free-run salmon and other wild fisheries;
- Depletion of topsoil and aquifers;
- Extraction of coal, silver, gold, chromium, molybdenum, bauxite, sulfur, other minerals; and
- Waste going to a landfill or incinerator (pay-as-you-throw). (Hawken et al., 1999)

What would *not* be taxed?

- Your paycheck (income taxes),

- Taxes for corporations,
- Taxes on savings, and
- Taxes on retirement.

Both individuals and businesses would be able to avoid taxes by changing their behavior, designs, processes, and purchases in order to support a carbon tax (Hawken et al., 1999). How does this form of taxation work in Europe? In Denmark, landfill taxes caused the reuse of construction debris to increase from 12% to 82% (20 times greater than the 4% average as seen in most industrial countries). In Holland, green taxes reduced the amount of heavy-metal leaks into lakes by 97% (Hawken et al., 1999). Other countries (Sweden, Britain, Germany, the Netherlands, and Norway) are implementing small trials in order to address environmental degradation and unemployment. Will the United States follow suit? Competitors' labor costs will be lowered, and innovation will be encouraged as Europe moves more towards tax shifting (Hawken et al., 1999). Do we want to be a #1 country, or be constantly just playing catch-up? Europe has jumped 30 years ahead of us in many environmental issues.

A traditional way to measure corporate performance against economic, social, and environmental parameters is called the Triple Bottom Line (TBL) (Figure 1.1). (Edwards, 2005; Edwards, 2010; Savitz, 2006). Savitz (2006) reports that a business ought to be able to measure and document a positive return on investment (ROI) on all three bottom lines – economic, environmental, and social (equity). This is also called people (social), planet (environmental), and profit (economic).

A company concerned about all aspects of the Triple Bottom Line is Interface Inc., a carpet manufacturing company. Its primary mission is to devise strategies for the company that support and restore natural ecosystems. Interface Inc. strives to show the entire industrial world by its deeds what sustainability is, and the range of topics it involves: people, process, product, place, and profits. Interface's sustainability goals for 2020 included becoming restorative to other companies through the power of influence (Anderson, 2006).

Interface Inc. saved over $231,000,000 by eliminating wastes; reduced the total amount of energy to make its carpets by 35%; increased renewable energy consumption by 12%; reduced the water needed by 78% in its modular carpet line; and reduced carbon dioxide emissions by 46%. In order to accomplish these extraordinary sustainability goals, a company needs to have the right people working for it. John Bradford, the Vice President of Manufacturing and Operations says,

> If you truly believe in the quest toward sustainability and if you believe our company can be the firm that starts the next industrial revolution, you will wholeheartedly join us. If you don't believe in or live up to this — if you are not a minister for sustainability — you probably don't belong here. (Doppelt, 2003)

These views are going to be the ones that all employers will have in the future, at least the companies that you will want to be associated with in your job.

In Kalundborg, Denmark, four companies mutually relocated to a shared location so that their wastes could be transferred to the other companies as material resources. Sludge from a fish farm was used as fertilizer on another farm. Sulfur dioxide from the power plant was mixed with calcium carbonate to produce gypsum for a plasterboard factory. Surplus steam from the power plant provided heat for 3,500 local homes (Edwards, 2010). This is an example of cradle-to-cradle, having no waste produced that is not used by another company as a resource and considered right from the beginning design stage.

As mentioned in the previous chapter, in order to avoid wastes and to produce environmentally safe fabrics, Designtex took a close look at the chemicals used in its processes, all 7,962 chemicals. Only 38 met their criteria and were selected as the basis with which to build an organic, environmentally friendly fabric line. The new line of fabrics became an award-winning financial success (Edwards, 2010). These are the products that informed customers want, and given a choice, are the products that they will buy. Manufacturing facilities around the world need to take notice.

Sustainability has to do with a whole cultural change. We have to keep moving toward sustainable choices. Any big leap encourages more steps, even small ones. Small steps maintain momentum until the budget permits a major step (Doppelt, 2003). Keep working on both the big and small steps because they all move us forward.

A mental alignment is also critical (Doppelt, 2003). That means to keep sending the same message, bombarding your employees and co-workers until they realize it is not possible to think or behave in an unsustainable way. This is important everywhere: at home, at work, on vacation. Who will do this for you, the individual? Keep sending the same message, keep the pressure on your family and friends until they realize it is not possible to think or behave in an unsustainable way. And we need to realize that it will take time for us all to develop new patterns and habits.

What might your life look like even next year? How much can be saved by eliminating wastes? How much of your energy consumption can be reduced? How can renewable energy be used instead? Can you reduce your water usage? How much can you to reduce your carbon dioxide emissions? Any initial success represents a starting point, not the end. Sustainability is a long-distance race. Every day you realize that, there is more you can do. You have to continually revisit your commitment to achieving sustainability and ensure it remains resolute.

Once the easy choices have been made, once the "low-hanging fruit" has been picked, efforts can plateau, and alignment beyond these plateaus must be revisited. Any company that takes the responsibility of sustainability seriously will be the big winner in the future (Doppelt, 2003). Any individual that takes the responsibility of sustainability seriously will also be the big winner in the future. Whoever you are, help everyone to rethink ... EVERYTHING!

References Cited for Chapter 13
Work

Anderson, Ray. (2006). Restoring the Planet, One Carpet Tile at a Time. Green Technology. Retrieved from http://www.green-technology.org/green_technology_magazine/ray_anderson.htm

Doppelt, Bob. (2003). *Leading Change Toward Sustainability: A Change-Management Guide for Business, Government and Civil Society*. Sheffield, UK: Greenleaf Publishing Limited.

Edwards, Andres. (2005). *The Sustainability Revolution: Portrait of a Paradigm Shift.* Gabriola Island, BC, Canada: New Society Publishers.

Edwards, Andres. (2010). *Thriving Beyond Sustainability: Pathways to a Resilient Society.* Gabriola Island, BC, Canada: New Society Publishers.

Flaherty, Joseph. (2009). *For the price of a TV you can start a FabLab*. Replicator. Retrieved from http://replicatorinc.com/blog/2009/10/for-the-price-of-a-tv-you-can-start-a-fablab/

Hawken, Paul, Amory Lovins, and L. Hunter Lovins. (1999). *Natural Capitalism*. New York, NY: Little Brown and Company.

Hightower, Jim. (2013). *You Won't Love These McSubsidies, Taxpayers will shell out $1.2 billion this year to support workers subjected to McDonald's miserly wages and benefits.* In Other Words. Retrieved from http://otherwords.org/wont-love-mcsubsidies/

Khanthong, Thanong. (2009). A revisit to Buddhist economics as a way out. The Nation. Retrieved from http://www.buddhistchannel.tv/index.php?id=8,7729,0,0,1,0#.UncbpSifu0s

Nathaniel, Jerome. (2011). Buddha v. Economics: David Loy applies Buddhist Principles to IMF and World Bank. Democrat and Chronicle. Retrieved from http://blogs.democratandchronicle.com/youngprofessionals/?p=2561

Savitz, Andrew. (2006). *The Triple Bottom Line: How Today's Best-Run Companies Are Achieving Economic, Social and Environmental Success -- and How You Can Too.* San Francisco, CA: Jossey-Bass, a Wiley Imprint.

Sivaraksa, Sulak. (2009). *The Wisdom of Sustainability: Buddhist Economics for the 21st Century*. Kihei, HI: Koa Books.

Steffen, Alex (editor). (2011). *World Changing: A User's Guide for the 21st Century*. New York, NY: Abrams Books.

Wolff, Richard. (2011). *The Truth About 'Class War' in America.* The Guardian. Retrieved from http://www.theguardian.com/commentisfree/cifamerica/2011/sep/19/class-war-america-republicans-rich

CHAPTER 14
RECREATION

Mark Anielski (2007) in his book *Economics of Happiness: Building Genuine Wealth* describes how to measure our progress: "Our wealth is total capital – social, natural, and financial" (Anielski, 2007). What are the things we need to protect in the country? What are the things we need to protect in Oregon?

The first thing we need to protect is our vacation time. As of 2011, the United States is the only nation, other than the Guianas, Nepal, and Myanmar, that does not have a paid vacation law (Steffen, 2011). More than 30% of Americans do not get any paid vacation, and more than half get one week or less. Because we have so little time off from work, we have no time to plan a sustainable vacation. We just need to get away.

Our recreational opportunities should help promote a healthier lifestyle. They should work to give us balance between the environment, and our social and economic conditions. Our recreational activities should involve community engagement, the respectful cultivation of relationships, and connect people to nature. Culture and nature should provide a range of recreational opportunities. All of these activities should teach us how to protect the natural, cultural, and scenic environment for present and future generations.

What do you do for recreation locally? Go to the movies? Do we have any sustainable options in our movies? Lights can be solar-powered, the camera too. Some theaters are going green! The Historic Palm Theater in San Luis Obispo, California, is the world's first fully solar-powered theater. It opened in 2004. The Palladio 16 Cinema in Folsom, California has the largest solar array of any cinema complex. Besides watching the movie, we also buy lots of snacks to eat during the movie. The Rerun in Brooklyn, New York, serves sustainable snacks and organic drinks. Is there anything to do about all the paper waste at the theater? Do you have an eco-friendly movie venue near you?

Beer is a basic staple for many football fans. Why not select one that takes the environment into account? Portland has many responsible brewers (more microbreweries per capita than any other city in the country). Some of these brewers make delicious ales from organic ingredients, get their energy from solar power, or use their spent grain to feed livestock. Find and support a local craft brewer.

What about taking a vacation *in* nature? Some use off-highway vehicles (OHV); others run, bike, and hike on the trails. Others prefer being near the water, where, of course, water quality will be of concern. And then others like the slopes and skiing. Each of these hobbies requires attention and care for the environment. Driving (alone) in a car is the most energy-intensive path to fun. Look for alternatives, such as public transportation or car-pooling.

Many of us enjoy getting away from it all by going out into nature. The effects of off-highway vehicles (OHV) and off-road vehicles (ORV) on soil, water, vegetation, heritage sites, and wildlife have long been recognized (Stokowski and LaPointe, 2000). To avoid damage to ecosystems, we must stay on maintained trails. Don't blaze trails through streams and wetlands; locate a bridge or don't cross. Then work with your Parks and Recreation representative to construct a bridge crossing where needed. Muddy and wet areas should be avoided, and bluffs also so that the soil structure is not destroyed. Where the ground is susceptible to rutting or erosion, these sensitive areas should be avoided, especially during spring melts. It takes over 200 years to rebuild soil structure once damaged (Figure 14.1).

Figure 14.1. Damage to a wetland marsh area due to illegal off-road vehicles (ORVs). Source: USDA, Forest Service.

Due to our presence in natural ecosystems, impact on some level is inevitable (Cole, 2004). We just have to decide what level of impact we are willing to accept. Impacts to natural systems occur instantly and recover very slowly. That means it is always better to prevent an impact than to have to repair damage. Damage to a new site is a much greater problem than additional deterioration of an established site (Cole, 2004). Damage occurs when we don't think about what we are doing (Figure 14.1). Damage occurs just from our presence (Figure 14.2).

Figure 14.2. Edge of Half Dome in Yosemite National Park, and the line of people waiting to hike the trail. Source: Wiki-Commons.

Many sports use specific clothing that we wear casually as well for their intended use. Do some research regarding the companies that manufacture the athletic apparel and equipment intended for these forms of recreation. How sustainable are they? For example, if you want eco-friendly, organic hiking boots, you should look for the following:

- Low-impact leather,
- Imitation leather (what kind is used?),
- Recycled rubber,
- Organic cotton or other organic fabrics,
- Hemp,
- Water-based adhesives and finishes,
- Organic dyes,
- Bees wax, and
- Hand-sewn products. (HubPages, Inc., 2014)

Look for either American-made products, or determine fair-trade information. The company Timberland is well known for its boots. In 2009, the company introduced a boot, Earthkeepers 2.0, intended to be recycled when no longer useful. Made of 80% recyclable materials, they were designed to be disassembled. Timberland recycles the soles and lining; the leather parts are cleaned, polished and readied to become a new pair of boots. Many additional styles have been added since the introduction of Earthkeepers.

In any form of activity, your impact within nature will be minimized if you:

1. Make careful plans and prepare ahead of time. Kindness to the earth is overlooked when we are rushed.
2. Whether on foot or by bike, travel on formal trails and camp in established sites to minimize disturbance to soils or damage to fragile flora and fauna. Respect wildlife. You are in their home.
3. Avoid campfires. Any open fire puts ecosystems at risk and are no longer necessary for comfort or food preparation.
4. Take your wastes home with you or dispose of them properly. Litter is ugly and can be deadly to other animals (Figure 14.3.). Over the period of a single month, two deer, a bear, and a fox were photographed because they got their heads caught in discarded containers (Tinker, 2013)! Two other bear cubs also got their heads stuck in a plastic cookie-jar at the time of the writing of this book (Kheiry , 2013; Rolando, 2014). How many animals do not find humans to free them from their traps, and die of dehydration and hunger?

5. Leave what you find in order to pass on the gift of discovery to others. It may be beautiful, but when you get home, how many seashells or feathers do you really need?

Many websites provide excellent guidance. For example, refer to the Leave No Trace Center for Outdoor Ethics (2014) for an extended list of recommendations.

Figure 14.3. Litter anywhere, along side a road or at a landfill, can be deadly for curious, hungry animals, such as this goat. Once trapped in the container, they can not eat or drink, and will die unless a human frees them. Source: Bigstock.com.

You also can do your part to keep natural areas beautiful by picking up litter that you find on the trails and by volunteering to maintain trails with local clubs. Although these are rules for hiking, similar rules exist for biking, and there is an environmental code of the slopes for skiers as well. Following appropriate codes protects and sustains our environment.

Some hiking sites are so popular that reservations may be required. The Half Dome hiking trail in Yosemite National Park is one of those popular destinations that requires permits to hike (Figure 14.2). A lottery takes place in the spring for the 254 hikers per day that are allowed (Half Dome, 2013). Hiking permits limit daily load and excessive wear on trails. There are some sites that even have a depart time assigned to maximize the natural experience and avoid excessive encounters with other hikers. For recreational sites along the water (beaches, rivers, and lakes), the main concerns regard water quality and cleanliness, as well as crowding. Just as on the trails, protecting aquatic environments and their vegetation is equally important. Never use chemicals near the water, not even soap or shampoo, which changes the pH of the water. Of course, swimmers should leave the water to use the bathroom. Urine is very high in nutrients and causes eutrophication (excessive algal growth).

Boaters, fishermen and fisherwomen should avoid spilling gas, oil, paint, varnish, or stripper in the water. All wastewater should be properly stored (both grey and human waste). Fish responsibly. That means do not dispose of fish entrails into public lakes, streams, or on shores. That can add unwanted nutrients to the water, can poison other animals, and can cause the area to smell bad for other visitors or residents.

What is the difference between ecotourism and sustainable tourism? Ecotourism refers to traveling to natural areas, whereas sustainable tourism applies social, environmental and economic consideration to all destinations (Steffen, 2011). When we think about sustainable vacations, we usually think about backpacking through virgin jungles or of snorkeling in pristine waters. But you should try to make your eco-footprint and cultural footprint small no matter where you go. Make businesses understand that you want to be a responsible tourist (Steffen, 2011). Fill out customer comment cards wherever you go and let resorts know you're paying attention to their green efforts or lack thereof (Ski Green, 2009).

Look for lodging that minimizes your impact on the environment, and use local transportation. Conserve water and energy by turning off lights when you leave the room, and re-use towels and linens (Ski Green, 2009). Take washable tableware and silverware to use in cafeterias instead of using paper and plastic utensils.

Take advantage of environmental educational programs wherever they are offered, to learn more about the local environment and how to protect it. For instance, there is a new movement for premium vacationers called Lifestyles of Health and Sustainability, or "LOHAS" (Travel Trends, 2011). LOHAS are people who are informed and aware, especially regarding social and environmental issues. They are looking for "tourism that is ecologically sustainable, and meets their standards of ethics and social justice" (Travel trends, 2011). They also will not tolerate "greenwashing." Greenwashing means a business spends more time and money claiming to be green, rather than actually changing their practices in a manner that truly considers the environment, and the culture upon which their company thrives.

If you must fly to your destination, use the Airbus A319, Boeing 787, or more fuel-efficient turboprop planes (Ski Green, 2009). Airlines are using more fuel-efficient planes, so this is an area to research before you finalize plans. You can also participate in carbon-offset programs to compensate for the carbon emitted by your flight. Carbon offsets fund programs whereby carbon dioxide emissions are reduced elsewhere, such as in the construction of wind farms, or in reforestation projects. By striving to become carbon-neutral yourself, you support clean technology, investment and innovation in a sustainable marketplace, and you hasten the transformation to a net-zero-carbon future. Many sites exist where you can calculate your carbon footprint (i.e. The Nature Conservancy, 2013) and provide places where you can offset your emissions (i.e. Terrapass, 2013). One thousand pounds (454 kg) of your carbon dioxide can be offset for less than $6. The fuel that is burned and the carbon dioxide that is released from your airline trip does not change. It is still released into the atmosphere. Your purchase of the carbon-offset sponsors a company that will, for example, plant trees to absorb an equal amount of carbon dioxide elsewhere. The total global carbon then either remains the same or improves, depending upon the specific program. This allows you to fly without contributing to global warming. The Carbon Neutral Company is one website that describes many carbon offset programs online (Carbon Neutral, 2014).

The National Ski Areas Association (NSAA) (2013) website lists a Green Room site where hundreds of improvements (both environmental and cost-saving) at resorts are summarized. This includes wind-powered ski lifts and buildings; snow-making equipment that is energy efficient or powered by micro-hydroelectric turbines; efficient lighting and heat re-use; biodiesel fuel from recycled cooking oil; composting; and toxic chemical avoidance.

If traveling to other countries, learn as much as possible about the religion, culture, and local values. Learn some of the language. Support locally owned businesses and local food (Intrepid's Top 12, n.d.). Travelling sustainably means valuing what is different about locations. Avoid food and hotel chains, because money spent there leaves the local economy and does not support the place you are visiting. However, do not patronize businesses that exploit or promote cruelty to endangered species. Bargaining, if a local practice, is fine; however, remember a small amount to you may represent a large portion of a seller's income (Intrepid's Top 12, n.d.). When photographing people, make it a 2-way exchange and offer to send copies to the individual. Instead of giving money to beggars, donate to a local development project that may be more supportive. Obviously, protect and respect the local environment.

In Oregon, we live in a beautiful state with mountains, ocean, desert, and river valleys—all unique places to visit locally. Without careful planning here, as well as around the world, it is inevitable that we will lose access to places due to negligence and carelessness. Recall my experiences described in Chapter One regarding the Pacific Northwest salmon runs; the direct access to the stones at Stonehenge, England; the right to see the ancient cave paintings in Lascaux, France. These are things that I have seen that you cannot (Figures 14.4 through 14.6).

Less destructive vacations require time to plan. To be caring enough to make good decisions requires time. Taking a train versus a plane requires time (Steffen, 2011). Think about doing whatever is necessary to protect the places you love. What, for your children, is at risk right now?

Figure 14.4. Plentiful salmon runs were commonplace in the Pacific Northwest in the 1960s. Source: Shutterstock.com.

Figure 14.5. Lascaux, France. Painted 17,300 years ago, the Lascaux caves now have a replica site that receives 730,000 visitors a year. Tourists are no longer allowed access to the original caves because body heat damages the paintings. Only serious scientists are allowed into the caves to see the original paintings. Source: Wikipedia User: Prof Saxx.

Figure 14.6. Stonehenge, England. Built in 2000-3000 BCE, the million visitors a year are no longer allowed within the stones (as shown above with the author's Father in 1962) but must now remain some distance away from the stones to prevent erosion. Source: (above) L. Pope; below: iStock.com.

References Cited for Chapter 14
Recreation

Anielski, Mark. (2007). *Economics of Happiness: Building Genuine Wealth.* Gabriola Island, BC, Canada: New Society Publishers.
Carbon Neutral Company,. (2014). Carbon offsets. Retrieved from http://www.carbonneutral.com/carbon-offsets

Cole, David N. (2004). Environmental impacts of outdoor recreation in wildlands. In: Manfredo, M.; Vaske, J., Field, D, Brown, P. and Bruyere, B. (eds). *Society and resource management: a summary of knowledge*. Modern Litho: Jefferson City, MO: 107-116

Kheiry, Leila. (2013) Ketchikan man saves bear cub that had head stuck in jar. *Ketchikan Daily News Fairbanks Daily News-Miner*. Retrieved from http://www.newsminer.com/ketchikan-man-saves-bear-cub-that-had-head-stuck-in/article_4415b125-0a83-5a17-a0bf-9353dc57ac3c.html

Tinker, Dani. (2013). Wildlife Are Getting Their Heads Stuck in Our Trash. *Wildlife Promise.* National Wildlife Federation. Retrieved from http://blog.nwf.org/2013/06/wildlife-are-getting-their-heads-stuck-in-our-trash/

Half Dome Day Hike Permits. (2014). Retrieved from http://www.nps.gov/yose/planyourvisit/hdpermits.htm

Hawken, Paul, Amory Lovins, and L. Hunter Lovins. (1999). *Natural Capitalism.* New York, NY: Little Brown and Company.

How to Ski Green. (2009). Retrieved from http://skigreenguide.com/?p=1148#more-1148

HubPages, Inc. (2014). Eco-Friendly, Organic Hiking Boots. Retrieved from http://hubpages.com/hub/Eco-Friendly-Hiking-Boots

Intrepid's Top 12 Responsible Travel Tips. (n.d.). Retrieved from http://www.intrepidtravel.com/sites/default/files/images/Intrepids_top_12_responsible_travel_tips.pdf

National Ski Areas Association (NSAA). Green Room. (2013). Retrieved from http://www.nsaa.org/environment/the-green-room/

(The) Nature Conservancy. (2013). Carbon Calculator. Retrieved from http://www.nature.org/greenliving/carboncalculator/

Rolando, Donna. (2014). Bear cub rescued after head gets stuck in cookie jar in Ringwood. Retrieved July 4, 2014, from http://www.northjersey.com/community-news/bear-cub-rescued-after-his-head-gets-stuck-in-cookie-jar-1.1043863#sthash.RBUQJCVB.dpuf

Steffen, Alex (editor). (2011). *World Changing: A User's Guide for the 21st Century*. New York, NY: Abrams Books.

Stokowski, P.A. and LaPointe, C.B. (2000). Environmental and Social Effects of ATVs and ORVs: An Annotated Bibliography and Research Assessment. November 2000. Burlington VT: School of Natural Resources, University of Vermont.

Terrapass. (2013). *Calculate your carbon footprint*. Retrieved from http://terrapass.com/calculate-carbon-footprint/?gclid=CLKA6Lyc27oCFUNBQgodulcAMw

Tinker, Dain. (2013). Wildlife Are Getting Their Heads Stuck in Our Trash. National Wildlife Federation. Retrieved from http://blog.nwf.org/2013/06/wildlife-are-getting-their-heads-stuck-in-our-trash/

Travel Trends: 2011 and Beyond. (2011). Retrieved from http://www.hoteliermiddleeast.com/10375-travel-trends-2011-and-beyond/#.Un7XVyifu0s

CHAPTER 15
HEALTH

Whereas other sources concentrate on the effects that the environment is having on us, I primarily want this chapter to have the opposite orientation. What effect do our health choices have on the environment? Can we look to nature, as in biomimicry, to find healthcare solutions? Could it be that if we bring health back to the planet that we will also become a healthier species at the same time?

According to Kreisberg (2004), there are ways we can live and die and still support the health of the ecosystems that sustain our lives. We just have to make a consciously informed decision every time we seek medical care. Ecologically Sustainable Medicine (ESM) or Green Health Care both represent a new health care model wherein the interconnectedness of humans and the planet are recognized.

A white man born in the United States in 1850 would not, on average, have been expected to live more than 40 years. The average lifespan for that same man today would be 75 years (Steffen, 2011). The increase is due to less violence, better nutrition, workplace safety laws and public-health medicine (Steffen, 2011). Although, due to the current decline in nutritional value of the food in our country, and the effect that genetically modified foods may be having on our health (potential for increased cancer rates), lifespans could begin to decline again.

Why are Americans so tall? We have had a good diet for several decades. Chemicals in the food we eat influence the gene mechanisms that control enzyme production. When we provide our genes with the right chemicals from healthy foods, it triggers enzyme production that results in growth spurts. Take that healthy diet away from the next generation and heights will shrink to previous averages (Benyus, 1997). Leaders of countries are usually taller than the average person. In the late 1700s to early 1800s, Napoleon, the first Emperor of France, at 5 ft 7 in. (1.70 m) was taller than average height. The average height for males in the United States currently is 5 ft 10 in. (1.78 m) (President Obama is 6 ft 1 in. [1.85 m]).

The environmental damage associated with food production parallels the damage to our health, with 34% of Americans falling into the category of being obese. Diabetes affects 8% of the population. Billions of dollars could be saved each year in medical costs if we had healthier diets (Edwards, 2010).

Global climate change is going to influence the number of epidemics around the world. How are diseases spread in our modern world of travel and the global transport of goods? How can we predict and prevent the spread of diseases? A curious collection of data regarding this mobility issue has recently become available from WheresGeorge.com. Enthusiasts track the location from where a dollar bill was entered into the system, and then follow the circulation of that bill across the country through the online site. If you are given a bill in your change that has "WheresGeorge.com" stamped on it, the hobbyist hopes you will go to the site, and record where you got the bill and where you spent it (Figure 15.1). The locations are mapped over time (Figure 15.2). For mathematicians, this has provided an enormous pool of data regarding travel and circulation patterns, and it has made it possible for them to build more accurate epidemic spread and response models from the data (Steffen, 2011).

Figure 15.1. Hobbyist's markings on dollar bills are used to track bill circulation. The data also gives mathematicians a way to track the spread of diseases and epidemics.

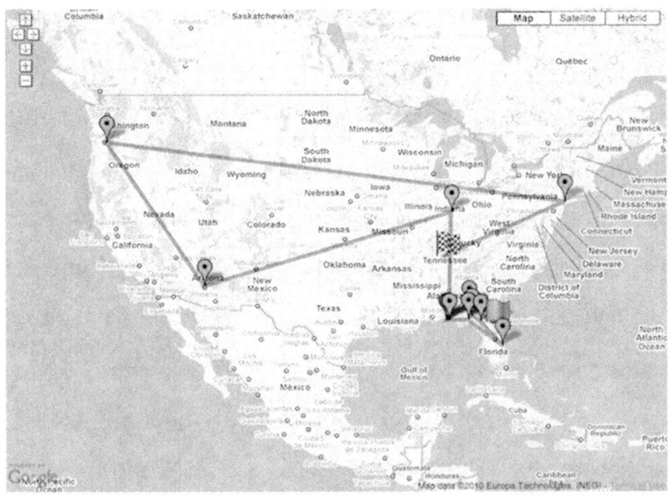

Figure 15.2. Tracking of a Where's George one-dollar bill released originally in Aiken, South Carolina, and its circulation across the country and through Portland, Oregon. Source: WheresGeorge.com

Turning again to the serious effect that we have on the environment by our health care choices, the issue of drugs in our water supply is of great concern. We rarely consider where most of our medicines end up. This includes acetaminophen and ibuprofen, as well as the residues of our excessive use of caffeine and nicotine (Donn et al., 2008). And we are consuming more and more drugs every year. According to Doheny (n.d.), drugs are commonly flushed down the toilet, as well as excreted from our bodies. As these drugs cannot be cleared from the water by our sewage treatment plants, they end up in our lakes and streams.

Two-thirds of our population is drinking water that has had fluoride added to it. Fluoride causes an extensive list of health problems, such as arthritis, cancer, diabetes, and infertility in men. Refer to the research articles that support these claims (Fluoride Action Network, 2012).

The problem of drugs in our water supply continues into natural settings. Once in nature, all sorts of forms of life consume the drugs when our contaminated water circulates back into the natural water supply. Wildlife, especially amphibians and fish that cannot escape the contaminants in the water, are being harmed in unexpected ways. For example, widespread deformities, like extra or missing limbs, are occurring in frogs exposed to these pharmaceuticals. Antibiotics, anti-depressants, birth-control pills, seizure medication, cancer treatments, painkillers, tranquilizers and cholesterol-lowering compounds have all been detected in our water sources. Anti-depressants have been blamed for altering the sperm levels in marine life and estrogen (female sex hormone used in birth control pills) deforms the reproductive systems of fish (Consumer Products, n.d.).

Daughton and Ternes (1999) reveal that the problem is just beginning:

> The enormous array of pharmaceuticals will continue to diversify and grow as the human genome is mapped. Today there are about 500 distinct biochemical receptors at which drugs are targeted. The number of targets is expected to increase 20-fold (yielding 3,000 to 10,000 drug targets) in the near future…This explosion in new drugs will severely exacerbate our limited knowledge of drugs in the environment and possibly increase the exposure/effects risks to nontarget organisms. (Daughton and Ternes, 1999)

This represents another reason why we should collect our own rainwater, as this is nature's way of purifying water. Rainwater is distilled, but not pure, so it still needs some treatment, but since it is distilled, the chemicals and pharmaceuticals are no longer present.

Dr. Masaru Emoto in his book *The True Power of Water* (2005) reminds us that we are 97% water ourselves. Although his work has been criticized for being unscientific, Emoto claims that our thoughts, music, and pollution all have an effect on our water. His studies show that pure water forms beautiful, perfect crystals, whereas polluted or even microwaved water lacks the ability to form a crystal at all. Either way, it gives us reason to think carefully about what we do to our water sources.

In the future, how might we actually heal ourselves? Using biomimicry, what can we learn from other mammals? Animals seem to know what they need and actually crave it. Benyus (1997) describes a famous study with rats that were given unlimited quantities of 11 different dishes containing protein, oils, fats, sugars, salt, yeast, etc., and the rats carefully selected a diet that enhanced their growth with the fewest calories. Somehow humans have lost this ability.

What do animals do when they are sick? As I walked my miniature poodle when he was sick, I watched him search for what seemed like hours for one particular grass. Benyus describes a chimp with a tummy ache that headed for an Aspilia plant, a plant with fuzzy leaves. After carefully rolling them around in his mouth, the chimp swallowed a dozen leaves whole. The fuzzy leaves were excreted intact, but tapeworm fragments were also excreted (Benyus, 1997).

Howler monkeys have absolutely no gum disease or tooth decay. It may have something to do with the fact that they eat huge quantities of cashew nuts. These nuts have high amounts of anacardic acid and cardol, both of which kill the bacteria that causes tooth decay in humans. Maybe a daily snack of cashews is all you need to prevent your next visit to the dentist (Benyus, 1997).

The Howlers also can decide whether they conceive male or female offspring. The X-chromosome is electropositive and the Y-chromosome is electronegative. Since like repels like, if the female Howler creates a negatively charged environment, positive sperm might be aided in their path to fertilization. Females decide whether it would be more beneficial to the Howler group to have a male or female offspring (Benyus, 1997).

In the Navajo language, many traditional medicinal herbs contain the name "bear." It is thought that the Navajo observed the bears self-medicating and then adopted their practices themselves (Benyus, 1997).

Plants make 500-600 compounds in each leaf, and each compound is involved in 50-60 different biological activities. The trick for science is to figure out which ones are performing all the miracles (Benyus, 1997). If you look at a typical synthetic vitamin pill taken by the average American versus one that is made from foods, the ingredient list is not at all comparable. The ingredients in a vitamin pill are there because we have studied those compounds. But there are hundreds of compounds present in food, and we have no understanding of their purpose. For example, there are over 200 nutrients in a carrot root (Figure 15.3) (Duke, 1992). Perhaps countries with vegetarian diets, or very little meat in their diets, are healthier than we are in the United States at this point.

2-METHOXY-3-SEC-BUTYL- PYRAZINE3,4-DIMETHOXY-ALLYL- BENZENE3-METHOXY-4,5-METHYLENE- DIOXY-PROPYL-BENZENE 5,7-DIHYDROXY-2-METHYL-CHROMONE 6-HYDROXY-MELLEIN 6-METHOXY-MELLEIN ACETALDEHYDE ACETONE ACETYLCHOLINE ALANINEBENZYLAMINE BERGAPTEN BETA-AMYRIN BETA-BISABOLENE BETA-CAROTENE BETA-CRYPTOXANTHIN BETA-FARNESENE BETA-PINENE BETA-SITOSTEROL BETAINEBIPHENYL BORNEOL BORNYL-ACETATE BORONDAUCOSTEROL DEC-2-EN-1-AL DECA-TRANS-2,TRANS-4-DIEN-1-AL DEHYDROASCORBIC-ACID DIOSGENIN DIPENTENE DODECAN-1-AL EO EPSILON-CAROTENE ETHANOL ETHYLAMINE ETHYL-METHYL-AMINE FALCARINDIOLISOCITRIC-ACID ISOLEUCINE ISOPIMPINELLIN ISOPRENE KAEMPFEROL-3-0-BETA-D-GLUCOSIDE KILOCALORIES LAURIC-ACID LECITHIN LEUCINE LIMONENE LINALOOL LINOLEIC-ACID LINOLENIC-ACIDNICKEL NITROGEN NON-2-EN-1-AL NONAN-1-AL NOPOL OCTAN-1-AL OLEIC-ACID OSTHOLE OXALIC-ACID OXYPEUCEDANIN P-COUMARIC-ACID P-CYMENE P-HYDROXYBENZOIC-ACID PALMITIC-ACIDSHIKIMIC-ACID SILICON SODIUM STARCH STEARIC-ACID STIGMASTEROL STRONTIUM SUBERIN SUCCINIC-ACID SUCROSE SULFUR SYRINGIC-ACID TARTARIC-ACID TERPINEN-4-OL - ALPHA-AMYRIN ALPHA-BERGAMOTENE ALPHA-CAROTENE ALPHA-CARYOPHYLLENE ALPHA-HUMULENE ALPHA-IONONE ALPHA-KETOGLUTARIC-ACID ALPHA-PHELLANDRENE ALPHA-PINENE ALPHA-TERPINENE ALPHA-TERPINEOL ALPHA-TOCOPHEROL ANILINEARABINOSIDE ARGININE ASCORBIC-ACID ASH ASPARTIC-ACID BARIUM BENZOIC-ACID-4-O-BETA-D-GLUCOSIDEBROMINE BUTYRIC-ACID CADMIUM CAFFEIC-ACID CAFFEOYLQUINIC-ACID CALCIUM CAMPESTEROL CARBOHYDRATES CAROTATOXIN CAROTOL CARYOPHYLLENE CARYOPHYLLENE-OXIDE CHLOROGENIC-ACID CHOLINECHROMIUM CIS-BETA-BERGAMOTENE CIS-GAMMA-BISABOLENE CITRIC-ACID COBALT COPPER COUMARIN CYANIDIN-DIGLYCOSIDE CYSTINE D-GLUCOSE DAUCIC-ACIDFALCARINOL FAT FERULIC-ACID FIBER FLUORINE FOLACIN FOLATE FRUCTOSE FUMARIC-ACID GALACTOSE GAMMA-BISABOLENE GAMMA-CAROTENE GAMMA-DECANOLACTONE GAMMA-MUUROLENE GAMMA-TERPINENE GERANIOL GLUTAMIC-ACID GLUTAMINEGLYCINE HCN HEPTAN-1-AL HERACLENIN HISTIDINE IONENEIRONLITHIUM LUPEOL LUTEIN LUTEOLIN-7-0-BETA-GLUCOSIDE LYCOPENELYSINE MAGNESIUM MALIC-ACID MALTOSE MALVIDIN-3,5-DIGLUCOSIDE MANGANESEMANNOSE METHIONINE METHYLAMINE MEVALONIC-ACID MOLYBDENUM MUFA MYRISTIC-ACID MYRISTICIN N-METHYL-ANILINE N-METHYL-BENZYLAMINE N-METHYL-PHENETHYLAMINE NEUROSPORENENIACIN (B)PALMITOLEIC-ACID PANTOTHENIC-ACID PECTIN PECTINESTERASE PEROXIDASE PHENYLALANINE PHOSPHOFRUCTOKINASE PHOSPHORUSPHYTIN PHYTOFLUENE PHYTOSTEROLS POTASSIUM PROLINE PROTEIN PSORALEN PUFA QUINIC-ACID RHAMNOSE RIBOFLAVIN (B) RUBIDIUM SABINENE SCOPOLETIN SELENIUM SERINE SFATERPINOLENE TETRADECENOIC-ACID THIAMIN (B) THREONINETIN TITANIUM TOLUIDENE TRANS-GAMMA-BISABOLENE TRYPTOPHAN TYROSINE URONIC-ACID VALINE VITAMIN A VITAMIN C VITAMIN B6 VITAMIN E VITAMIN K WATER XANTHOPHYLLS XANTHOTOXIN XYLITOL XYLOSE ZINC ZIRCONIUM

Figure 15.3. More than 200 nutrients are found in the carrot root. Source: Duke (1992).

When crops fail in Africa due to drought, the native tribes watch animals in order to determine what they can eat. Even the U.S. Navy understands that animals may hold the clues to survival. In the 1943 book, *How to Survive on Land and Sea*, the United States Office of

Naval Operations says "It is safe to try foods that you observe being eaten by birds and mammals" (U.S. Office of Naval Operations, 1943).

We often think we are superior to other animals. For many, it is embarrassing to think we need to learn from these "lower' animals. We have only recently decided that local indigenous cultures may have primitive knowledge of high value. If we can include animals in our knowledge base, and it is not too late, that may be another source of valuable information (Benyus, 1997).

Indigenous people have known forever that plants support each other and keep each other healthy. Wild plants around the edges of fields keep the plants in the field vital and healthy too. Corporate farming practices have contributed to the elimination of small farms, and therefore the hedgerows that used to separate them. This has eliminated an important part of the ecosystem, as the edge plants provided vitality to the crops.

All plant communities contain a few extremely rare plants in small numbers. Their purpose may seem inconsequential, but all plants contain unique and sometimes highly potent chemicals. The greater the variety of plants with diverse chemistries, the more vital and healthy the ecosystem. One small, rare plant contains special pharmaceuticals for conditions that may only occur in the ecosystem once in a century (Buhner, 2002). This should change our view of what it means to preserve biodiversity in all areas. No organism is dispensable.

There is evidence saying that if our environments are healthy, we will have healthy individuals. Living sustainably means we can create both healthy environments and individuals as we change the world and begin to live in harmony with the Earth. Consider, however, how we treat our plants, even in the city. How long could you live in solitary confinement (Figure 15.4)? Trees like to be in physical contact with each other, with roots touching. A common street tree in the city will live 10-15 years, whereas the same tree in a rural setting may live over 300 years. That may have something to do with how we treat our urban trees.

Figure 15.4. "Solitary confinement" of trees in urban environments, with restricted root growth, limited access to water and nutrients, resulting in short lifespans (300 years in a forest, 15 years in the city). Source: iStock.com.

Again, around the world we will solve problems differently in developed versus developing countries. In many countries around the world, opticians are hard to find, and getting good eye care is challenging. Joshua Silver (2009) worked to address these issues. He designed self-adjusting spectacles, which the wearer can easily adjust to his or her own prescription. Silver's goal is to distribute a billion pair to the world's poor by 2020. By 2011, he had already distributed 30,000 pair in 15 countries.

How can *your* choices for healthcare be made more sustainably?

References Cited for Chapter 15
Health

Buhner, Stephen H. (2002). *The Lost Language of Plants.* White River Junction, VT: Chelsea Green Publishing.

Consumer Products Threaten Aquatic Life: What Citizens Can Do. (n.d.). Sierra Club. Retrieved from http://www.sierraclub.org/toxics/downloads/fs4.pdf

Daughton, Christian and Thomas A. Ternes. (1999). *Pharmaceuticals and Personal Care Products in the Environment: Agents of Subtle Change?* Environmental Health Perspectives. Vol 107, Supplement 6. December 1999. Retrieved from http://www.epa.gov/esd/bios/daughton/errata.pdf

Doheny, Kathleen. (n.d.). *Drugs in Our Drinking Water? Experts put potential risks in perspective after a report that drugs are in the water supply.* From WebMD, Retrieved from http://www.webmd.com/a-to-z-guides/features/drugs-in-our-drinking-water

Donn, Jeff, Martha Mendoza and Justin Pritchard. (2008). *Drugs Found in Drinking Water*. Associated Press. Retrieved from http://usatoday30.usatoday.com/news/nation/2008-03-10-drugs-tap-water_N.htm

Duke, James A. (1992). Handbook of Phytochemical Constituents of GRAS Herbs and Other Economic Plants. Boca Raton, FL: CRC Press LLC.

Emoto, Masaru. (2005). *The True Power of Water: Healing and Discovering Ourselves.* Hillsboro, OR: Beyond Words Publishing Inc.

Fluoride Action Network. (2012). Health Effects. Retrieved from http://fluoridealert.org/issues/health/

Kreisberg, Joel. (2004). "What is 'Sustainable' Health Care?" *Journal of Ecologically Sustainable Medicine*. 2/3

Silver, Joshua. (2009). Adjustable liquid-filled eyeglasses. TEDGlobal. Retrieved from https://www.ted.com/talks/josh_silver_demos_adjustable_liquid_filled_eyeglasses

Steffen, Alex. (editor). (2011). *World Changing: A User's Guide for the 21st Century.* Abrams, NY: Sagmeister Inc.

United States, Office of Naval Operations. (1943). *How to Survive on Land and Sea*. Annapolis, MD: United States Naval Institute.

CHAPTER 16
THE WORLD

> "Everything on the earth has a purpose, every disease an herb to cure it, and every person a mission. This is the Indian theory of existence." — Mourning Dove [Christine Quintasket] (1888-1936), Salish Tribe. (Legends of America, 2013)

If we are going to have an amazing life, we have to think about the kind of world we want to live in. We have to think about our place in the country and in the world (Figure 16.1 and 16.2).

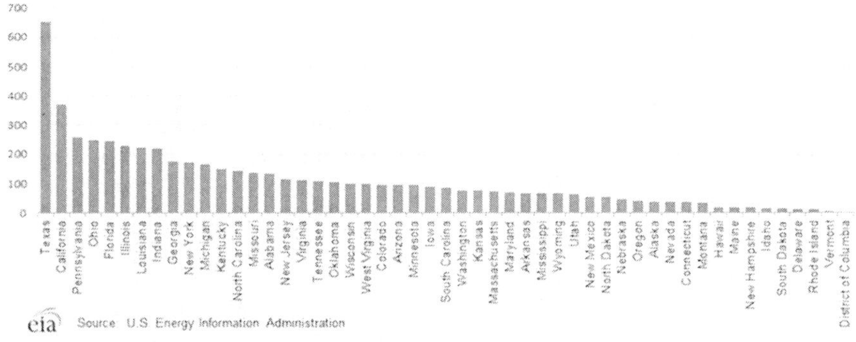

Figure 16.1. Energy-related emissions by state (in million metric tons of carbon dioxide), 2010. Source: EIA.

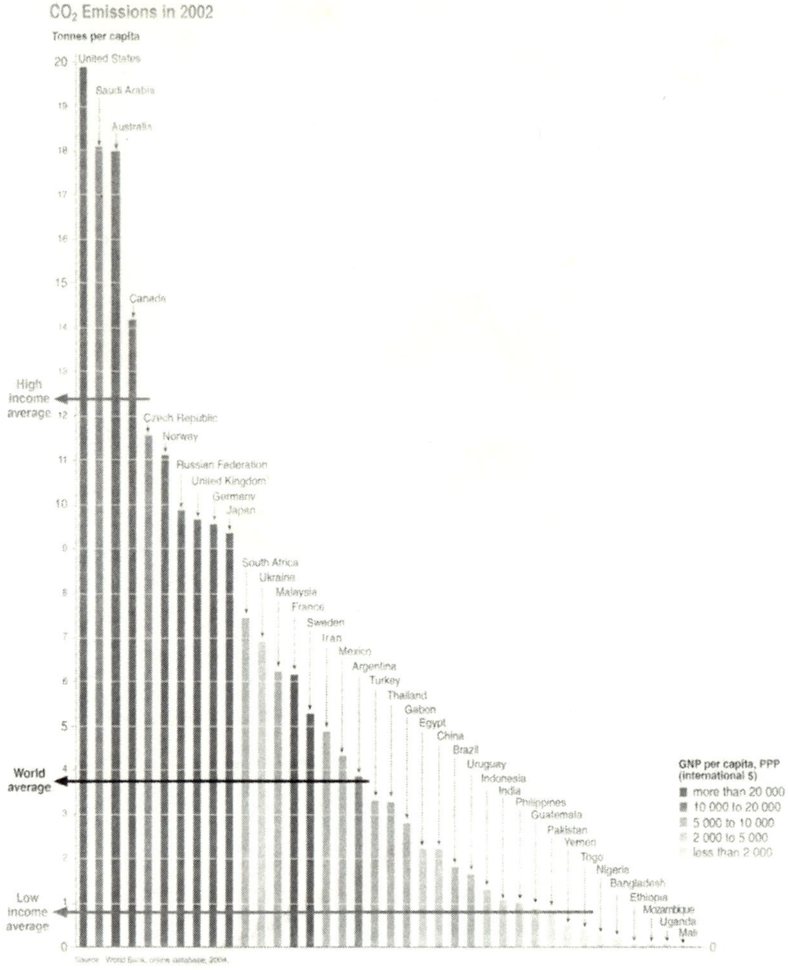

Figure 16.2. National carbon dioxide (CO_2) emissions per capita in 2002 (2005). Source: UNEP/GRID-Arendal Maps and Graphics Library.

Unlike the United States, Sweden has set a goal to be the world's first oil-free nation by 2020 (Edwards, 2010), increasing their use of renewable energy by 50%, making their vehicles fossil-fuel free by 2030, and having a net-zero greenhouse gas emissions by 2050. They reduced their greenhouse gas emissions by almost 10% from 1990-2010, while increasing their GDP by 48% (Edwards, 2010).

Four countries are in the race to be the first carbon-neutral (zero-carbon) country: Iceland, New Zealand, Norway, and Costa Rica (Lean and Kay, 2008). According to the authors, Iceland has gone the furthest, with only 1% of its homes heated with fossil fuels (all others are geothermal and hydroelectric). The Icelanders' love of big cars will be much more of a challenge to address. New Zealand plans to generate 90% of its energy from renewable sources by 2025, with agriculture releasing 50% of its emissions being its most difficult problem. Norway gets 95% of its power from hydroelectric sources, and heavily taxes cars and fuel. Costa Rica plans to soak up emissions by planting trees; they planted a world record of 5 million trees in a single year (2007), and their banana industry also plans to go carbon neutral. Their expansion in transportation increases (planes and cars) in the last decade will make their goal challenging. We each have areas where we excel and other areas that challenge us, and it is different for each country and community, and person.

Tocco, Italy, is a carbon-neutral city. The entire city, all 2,700 inhabitants, get 100% of their energy from 4 wind turbines (Schwartz, 2010). These turbines generate 30% more energy than they need, providing $200,000 a year for the city. This has allowed the city to eliminate local taxes, fees for garbage collection and street cleaning, and permitted the retrofitting of the local school for earthquake protection.

At Castello Monte Vibiano Vecchio, in central Umbria, Italy, the owners of this grape and olive farm north of Rome have the goal of being the first zero-carbon farm without carbon offsets. Twenty-four solar panels power the farm, and miniature electric tractors are powered by biofuels. When the trees are thinned, the wood chips are used in the olive oil boilers (Maeztri, 2008; Edwards, 2010).

Hong Kong unveiled its first zero-carbon building. The new building hopes to raise community awareness of sustainable living in Hong Kong. Some of the key features of the site include: carbon-neutrality generating more energy on-site than is used; the first urban native woodland in Hong Kong; and solar panels that completely cover the roof (First zero-carbon, 2012). However, the building is seen primarily as educational or inspirational. (Ko, 2012). This showcase is only 3 stories tall, while most buildings in Hong Kong are high-rises with high energy requirements. Zero carbon will not be easy for Hong Kong to achieve (Ko, 2012). While the European Union countries have mandated that all new homes must have zero-carbon emissions (by 2016 for the U.K.; 2020 for Denmark, etc.), these goals are seen as impossible in Hong Kong (Zero Carbon, 2009).

Kerala, India, is about one-sixth the size of Oregon, with 8 times more people. It has low consumption rates, while having a high quality of life. Life expectancy is age 72, with low infant mortality rates (lower than the U.S. or Sweden) (Edwards, 2005). It has 90% literacy, and approaches a Western standard of living with a gross domestic per capita income of only $1,000 per year (Edwards, 2005). A high quality of life is not dependent upon material wealth. Using this understanding, how might we improve the quality of life in our community?

How about the United States? How about Oregon? Portland? In the U.S., most efforts are at the city level. The International Council of Local Environmental Initiatives (ICLEI) is a global movement of more than 1,000 cities, villages, and towns that include more than

533 cities and counties in the United States (ICLEI, 2013). Recognized as a leader on climate action and clean energy, ICLEI sets national standards and provides information and trainings to support efforts to advance city goals. The ICLEI Portland, Oregon (2013), goals for the global warming initiative include:

- saving residents and businesses money by improving energy efficiency;
- making it convenient to walk or bicycle, and providing public transportation that is convenient, quick, and affordable;
- stimulating development of new, renewable energy resources;
- reducing solid waste generated and increasing recycling;
- accelerating urban re-forestation;
- improving understanding of climate change; and
- fostering a vibrant, livable community. (ICLEI USA, 2013)

The future described by this initiative is also summarized by Edwards (2010) in the SPIRALS framework. Edwards describes the SPIRALS approach through "thriveable projects, which are Scalable, Place-making, Intergenerational, Resilient, Accessible, Life-affirming, and involve Self-care" (2010).

Scalable: Projects that can easily be scaled from a community or small town to the level of a city or beyond (Edwards, 2010).

Place-making: These initiatives "consider the historical, cultural and biological characteristics of a community, its settlement history, values, natural cycles, environmental impacts and present concerns" (Edwards, 2010) (refer to Chapter 10).

Intergenerational: Focus is on the long-term benefits. An example of this intergenerational viewpoint is the sustainable forestry practices of the Menominee Nation in northeast Wisconsin. This Native American tribe has been practicing sustainable forestry for 150 years (Menominee, 1997). Thirty million board feet of timber is removed from the forest each year with unnoticeable changes to the landscape. The Menominee take inventory of the forest every 10-15 years to monitor changes. Each of the 58,000 trees is numbered, and their diameter and height measured. This aids the foresters in maintaining diversity, quality and quantity so that their forest always retains a mix of species and a variety of tree ages, including old growth white pine and sugar maple several hundred years old (Johnson and Johnson, 2012).

A video trailer for the documentary *For the Next Seven Generations* describes the work of the International Council of Thirteen Indigenous Grandmothers. The council includes grandmothers, wise elders, shamans and medicine women, from all over the world who have come together to help heal the world (For the Next Seven, 2013). This is an example of intergenerational consideration.

Resilient: If any system is under stress, its ability to bounce back from some shock, whether it is environmental or economic, is difficult. Dodman (2009) writes it may be more useful to instead be "bouncing forward to a state where shocks and stresses can be dealt with more efficiently and successfully" (Dodman, 2009). Change is inevitable, and adapting is key to surviving. Communities show their strength when they are adaptable, flexible, and resilient (Dodman, 2009; Edwards, 2010).

Accessibility: Communities that include all of their inhabitants in decision making, focus "on developing tools and job opportunities that incorporate aspects of the social and environmental justice" (Edwards, 2010) as well. They are inclusive of all residents; this means that everyone has equal or full access to everything in the community. Accessibility includes those with disabilities or other special needs. Universal design, or inclusive design, includes everyone, whether a person has a disability or not (Burgstahler, 2012). The late Ronald L. Mace encouraged design to "be aesthetic and usable to the greatest extent possible by everyone, regardless of their age, ability, or status in life" (Mace, 1998).

In Sonoma County, California, a web-based social marketing tool called Community Pulse helps to educate residents regarding their ecological footprint (Edwards, 2010). It tracks CO_2 emissions, energy and water usage, as well as the waste generated at the individual, county and state levels (Community, n.d.).

Life-affirming: These initiatives support the views described by Janine Benyus to use biomimicry and learn to copy nature. With a task in hand, if we ask how nature would solve this problem, we begin to understand the potential impact that our actions have (Edwards, 2010). If we follow the "Precautionary Principle," where "the key element of the principle is that it incites us to take anticipatory action in the absence of scientific certainty," (SEHN, 2014), we will at least do no additional harm. In essence, this principle places the burden of proof on a business to show that an action is safe, rather than on the public to prove it is harmful (Edwards, 2010). This life-affirming approach anticipates potential harm from an action and prevents it before it is too late and damage has already taken place. There is a popular quote (sometimes attributed to the anthropologist Jane Goodall), that goes something like, "Only if we understand, can we care; only if we care, will we help; only if we help, shall we all be saved." This life-affirming perspective explains how we can move toward a thriveable future (Edwards, 2010).

Self-care: This is what is needed if we are to be sustained through this age of change, as we do our best to make a difference. We will each have different ways to express this, through art or music, a hike in the woods, a stroll at sunset on the beach. "Our work is most effective when we replenish ourselves, and then give to others" (Edwards, 2010). Participation in the many environmental organizations, such as Green Drinks or EcoTuesday, provide social gathering opportunities where replenishment can occur and inspiration is provided between members.

Our main problem is that "there is no user's manual for how to live and operate on earth, the most important and complex system known" (Hawken et al., 1999). All people everywhere have similar wishes for the future of their children or grandchildren. "They want better schools, a better environment, safer communities, family-wage jobs, more economic security, stronger family support, lower taxes, more effective governments, and more local control" (Hawken et al., 1999). In this we are all the same. A clear starting point may be called restraint, the conscious use of resources, a powerful practice that is needed at this point in history (Benyus, 1997), placing value on what we have left.

No matter how much we may want it, we cannot at this time buy a better future. It is not available for sale at any price (Steffen, 2011). And all of the "green products around us are at best an advertisement for the idea that we should live sustainably, a sort of 'shopping therapy for the ecologically guilty'" (Steffen, 2011). We have an idea of sustainability as it starts to form in our daily lives, but an understanding of what it might be and actually living it are not the same. We need millions of people working in their areas of expertise to make every aspect of business and life sustainable, to be as socially responsible as possible (Steffen, 2011). It does not matter what your profession is; we need everyone to envision what their job might be in a fossil fuel-free, zero-waste, zero-water, zero-carbon, zero-energy world, and begin to work seriously in that direction.

One place for an individual to start is the National Geographic Quizzes by National Geographic (2013). Food, water, energy, and travel represent the topics of 19 quizzes available on their website, that test your understanding and show you how to reduce the impact of your environmental footprint on the planet (Green Quizzes, 2013).

So what should we have for goals? Who will be the first to reach zero-carbon? Zero-water? What should we have for our goals? General Electric says the Net-Zero Energy home was possible by 2015. Of course, instead of the net-zero energy home, the goal should be a thrivable one: a home that generates more electricity than it uses. Thrivable means to not only have no impact on the earth by our actions, but to restore the earth as well (Edwards, 2010).

When you start, begin by making a list of new ideas, a positive list. Don't just substitute ingredients in our new world; throw the recipe out the window, and start from scratch! Don't just reinvent the recipe; rethink the whole menu (McDonough and Braungart, 2002). Be ready to take your ideas further; no matter how good your idea is, it is not necessarily the best you can do. Keep exploring!

Lester R. Brown writes in his book *Plan B 4.0*, "Saving civilization is not a spectator sport" (Brown, 2009). He lists four things to do: educate yourself about the environmental issues; spread the word by writing letters to the editor, and on the Internet; get politically involved and let elected officials know what you think is important; and finally, take action in any area that you find to be exciting. He ends the book with the following:

> The choice is ours—yours and mine. We can stay with business as usual and preside over an economy that continues to destroy its natural support systems until it destroys itself, or we can adopt Plan B and be the generation that changes direction, moving the world onto a path of sustained progress. The choice will be made by our generation, but it will affect life on earth for all generations to come. (Brown, 2009)

Radical changes for the better are possible. All great movements for social change begin with statements of great optimism (Steffen, 2011). When implementing change, the involved members say, "It feels good to be part of the solution and not only the problem" (Walljasper, 2007).

Begin at the local level. Start in your own home, in your neighborhood, within your city, and at the state level. Realize that you can also work at the global level. Bill McKibben travels around the country inspiring others to join him in the fight to reduce carbon dioxide levels. He started the organization 350.org to address this issue globally as described in Chapter 9. He has given us three numbers to remember:

> Two; 565; and 2,795. "Two" represents the maximum increase in degrees Celsius for the global temperature; scientists everywhere have agreed on this limit. The number 565 equates to the maximum release in gigatons of carbon dioxide that can be released and still stay below the 2°C of global temperature increase. And the final number, 2,795 represents the gigatons of carbon dioxide currently in the fossil fuel industry fuel reserves, or five times the amount that we can safely burn and not lose control of the weather. (Fleischer, 2012)

Even though it is an enormous undertaking, this single person, Bill McKibben is working seriously to get us all to change and bring CO_2 emissions down to 350 ppm. His global team "helped to mobilize over 5200 actions in 181 countries in 2009, the most widespread day of political action in the planet's history" (350.org, n.d.). The largest global climate march in history was planned for September 21, 2014. We can make a difference!

There is good news out there. Unfortunately, you have to really do your research and dig deep, as strange as that may seem. Why should it be hidden from us, or not reported to the general public? For example, Hausfather (2013) reports that the United States' CO_2 emissions are below 1996 levels. This is great news! I can only imagine why this *good* news should not be reported. The small actions of multiple industries and those of individuals are adding up (Figure 16.3). We *do* make a difference! It is only with our continued efforts that we can reverse the damage done to our ecosystems, and to each other. Another world is possible!

Individual sustainability is just the beginning for the next challenge: sustainability at the neighborhood scale. Once we realize that we can make a difference at the global scale, and we can make a difference at the individual scale, then the scale at which we can without a doubt make a huge difference is that within our own neighborhoods. We don't have to tackle the whole world. And we don't have to solve all the issues alone. In fact, it will happen much faster when we work in small groups in which we can support each other. A group of people, wherein we can recognize all the members, and even know most by name, represents a group with enormous potential. This is the potential that awaits us. I hope you will realize that it is rewarding to be a leader in your own neighborhood. I sincerely wish you luck on this journey.

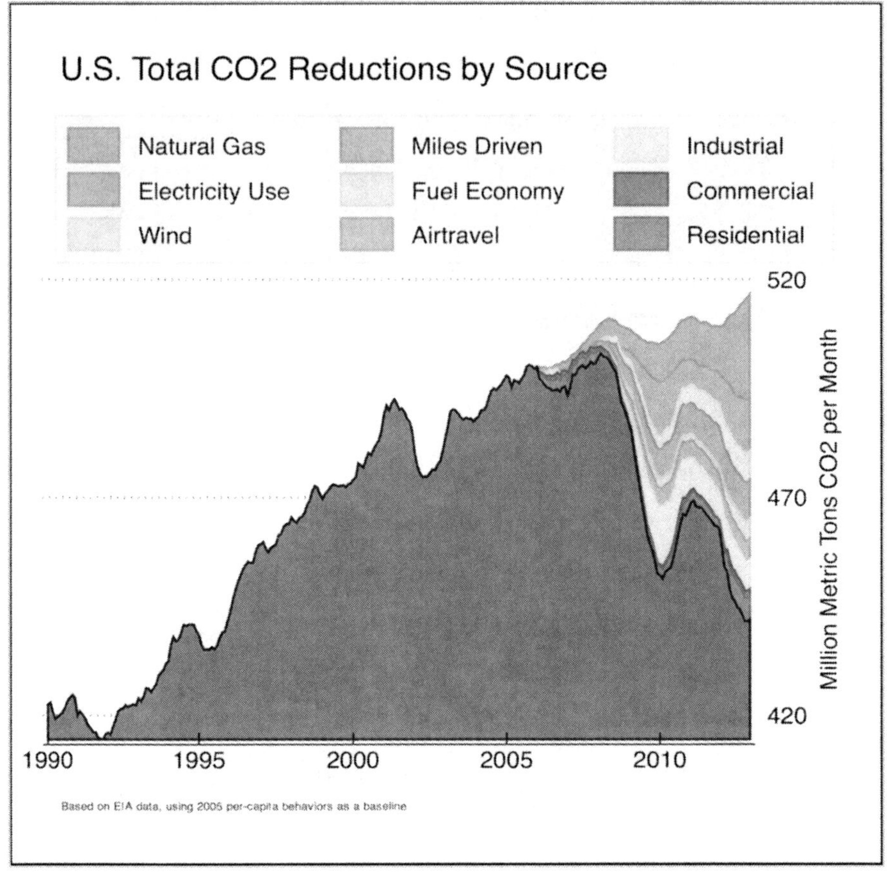

Figure 16.3. "What's Behind the Good News Declines in U.S. CO^2 Emissions?" by Zeke Hausfather (2013) and Hausfather (n.d.). Source: EIA. Refer to the original articles for color differentiation between sectors.

References Cited for Chapter 16
The World

350.org. (n.d.). Our Team's History. Retrieved from http://350.org/story

Benyus, Janine. (1997). *Biomimicry: Innovation Inspired by Nature*. New York, NY: HarperCollins Publishers Inc.

Brown, Lester R. (2009). *Plan B 4.0, Mobilization to Save Civilization*. Earth Policy Institute. New York, NY: W. W. Norton & Company. Retrieved from http://www.earth-policy.org/images/uploads/book_files/pb4book.pdf

Burgstahler, Sheryl Ph.D. (2012). *Equal Access: Universal Design of Your Project: A checklist for making projects welcoming, accessible, and usable*. Retrieved from http://www.washington.edu/doit/Brochures/PDF/design.pdf

Carbon Footprint. (2010). CO2 emissions in 2002. Retrieved from http://www.eoearth.org/view/article/150926/

Community Pulse. (n.d.) Providing Indicators for Action. Retrieved from http://www.communitypulse.org/

Diamond, Juli. (2013). Sweden Prepares to Lead EU on Climate. Worldwatch Institute. Retrieved from http://www.worldwatch.org/node/6039

Dodman, David. (2009). Building Resilience. State of the World, 2009. Worldwatch Institute. Retrieved from http://www.worldwatch.org/files/pdf/SOW09_chap5.pdf

Edwards, Andres (2010). *Thriving Beyond Sustainability: Pathways to a Resilient Society.* Gabriola Island, BC, Canada: New Society Publishers.

(The) First zero carbon building in Hong Kong. (2012). Building Journal. Hong Kong. Retrieved from http://www.building.com.hk/feature/2012_0813zcb.pdf

Fleischer, Matthew. (2012). Bill McKibben's 'Do The Math' tour points to 2028 as the year of catastrophic climate change. Huff Post Green. Retrieved from http://www.huffingtonpost.com/2012/11/13/bill-mckibben-do-the-math-2028_n_2123406.html?view=print&comm_ref=false

For the Next Seven Generations. (2013). Laughing Willow. Retrieved from http://vimeo.com/6538094#

Green Quizzes. (2013). National Geographic. Retrieved from http://environment.nationalgeographic.com/environment/green-guide/quizzes/

Hausfather, Zeke. (2013). What's Behind the 'Good News' Declines in U.S. CO_2 Emissions? The Yale Forum on Climate Change and the Media. Retrieved from http://www.yaleclimateconnections.org/2013/05/whats-behind-the-good-news-declines-in-u-s-co2-emissions/

Hausfather, Zeke. (n.d.). Explaining and Understanding Declines in U.S. CO_2 Emissions. Retrieved from http://static.berkeleyearth.org/memos/explaining-declines-in-us-carbon.pdf

Hawken, Paul, Amory Lovins, and L. Hunter Lovins. (1999). *Natural Capitalism*. New York, NY: Little Brown and Company.

ICLEI. (2013). Local Governments for Sustainability USA. Retrieved from http://www.icleiusa.org/about-iclei/members

ICLEI USA. (2013). Portland City Council Unanimously Adopts the Portland Plan. Bureau of Planning and Sustainability. Retrieved from http://www.icleiusa.org/news/portland-city-council-unanimously-adopts-the-portland-plan

Johnson, Christopher and Barbara Johnson. (2012). Menominee Forest Keepers. American Forests: Protecting and Restoring Forests. Retrieved from http://www.americanforests.org/magazine/article/menominee-forest-keepers/

Ko, Vanessa. (2012). *Hong Kong unveils its first zero-carbon building*. Global Observer. Retrieved from http://www.smartplanet.com/blog/global-observer/hong-kong-unveils-its-first-zero-carbon-building/

Lean, Geoffrey and Bryan Kay. (2008). *Four Nations in Race to Be First to Go Carbon Neutral: Iceland, New Zealand, Norway and Costa Rica are all hoping to turn their economies green, but the challenges they face are formidable.* The Independent. Retrieved from http://www.independent.co.uk/environment/climate-change/four-nations-in-race-to-be-first-to-go-carbon-neutral-802627.html

Legends of America. (2013). *Native American Legends: Great Words from Great Americans*. Retrieved from http://www.legendsofamerica.com/na-quotes.html

Lerch, Daniel. (n.d.). Share-It-Square. Project of Public Space. Retrieved from http://www.pps.org/great_public_spaces/one?public_place_id=505

Mace, Ronald L. (1998). A perspective on universal design. The Center for Universal Design. Environments and Products for All People. Retrieved from http://www.ncsu.edu/ncsu/design/cud/about_us/usronmacespeech.htm

Maeztri, Devin. (2008). Italy aims for carbon-neutral farm. Sustainable Cities Net. Retrieved from http://www.sustainablecitiesnet.com/models/italy-aims-for-carbon-neutral-farm/

McDonough, William and Michael Braungart. (2002). *Cradle to Cradle: Remaking the Way We Make Things.* New York, NY: North Point Press.

Menominee Forest-Based Sustainable Development Tradition. (1997). Retrieved from http://www.epa.gov/ecopage/upland/menominee/forestkeepers.pdf

Schwartz, Ariel. (2010). *Ancient Italian Town Completely Powered by Renewable Energy*. in Inhabitat – Sustainable Design Innovation. Eco Architecture. Green Building. Retrieved from http://inhabitat.com/ancient-italian-town-completely-powered-by-renewable-energy/

SEHN. (2014). Wingspread Conference on the Precautionary Principle. Retrieved from http://www.sehn.org/wing.html

Steffen, Alex. (editor). (2011). *World Changing: A User's Guide for the 21st Century*. Abrams, NY: Sagmeister Inc.

Walljasper, Jay. (2007). *The Great Neighborhood Book: A Do-It-Yourself Guide to Placemaking*. Gabriola Island, BC, Canada: New Society Publishers.

Zero Carbon Emissions. (2009). *Homes Go From 'Superefficient' to Zero Carbon Emissions in Europe*. New York Times. Retrieved from http://www.nytimes.com/cwire/2009/08/10/10climatewire-homes-go-from-superefficient-to-zero-carbon-e-3531.html?pagewanted=print

INDEX

Note: Page numbers with an italic "*f*" indicate charts, graphs, photographs, and maps

A

B

C

D

E

F

G

H

I

J

K

L

M

N

O

P

R

S

T

U

V

W

ABOUT THE AUTHOR

Linda C. Pope has been teaching sustainability, environmental science, and other life sciences at the college level for nearly twenty years. She has also taught special programs in earth science for kindergarten through fifth grade. She holds a Masters in Landscape Architecture from Harvard University and a Masters in Science (Plant Physiological Ecology) from the University of Maryland. She developed the Individual Sustainability course for Portland Community College (Oregon). She has lived all over the United States, as well as in England, France, and Algeria. Her broad base of experience and her commitment to effective community action provide a strong foundation from which to present the nearly infinite opportunities for positive change in the field of sustainability.

CPSIA information can be obtained
at www.ICGtesting.com
Printed in the USA
LVOW07s1626200817
545700LV00001B/146/P